"This is an excellent book, and it provides a helpful survey of the area while simultaneously exploring many arguments and ideas in significant depth. Arnold, Brennan, Chappell, and Davis provide a good diversity of perspectives on beneficence—Effective Altruism in particular—and when it is good, just, virtuous, and morally required. In each of the four parts, the authors collectively demonstrate that the issues discussed are rich and multidimensional; there are often more than just two sides to take (sometimes there are at least four!)."

Theron Pummer, *University of Saint Andrews, UK*

QUESTIONING BENEFICENCE

Effective Altruism is a movement and a philosophy that has reinvigorated the debate about the nature of beneficence. At base, it is the consistent application of microeconomic principles to beneficent action. The movement has exposed that many forms of giving do little good (or do active harm), but others do tremendous good.

Questioning Beneficence uses Effective Altruism as a launch pad to ask hard questions about beneficence more generally. Must we be Effective Altruists, or are Effective Altruism and the ideas driving the movement a mistake? How much should we give —if anything—and how should we give it? What are the respective roles of different kinds of institutions? Is charity anti-democratic and do billionaire philanthropists have too much power? Is Effective Altruism just utilitarianism in disguise?

Questioning Beneficence is written by four philosophers, each with distinct points of view. It introduces a new standard for debating ideas in philosophy as each author poses and answers three questions and each of his three co-authors responds to those questions in turn. Finally, the first author replies to his co-authors' responses. Throughout the book, there is a spirit of curiosity, intellectual risk taking, and truth-seeking, rather than point-scoring and one-upmanship. This book demonstrates what open-minded, real dialogue on an important issue can be at its very best.

Samuel Arnold is an Associate Professor of Political Science at Texas Christian University, USA. His research interests include liberalism, economic justice, and alternatives to capitalism.

Jason F. Brennan is Robert J. and Elizabeth Flanagan Family Professor of Strategy, Economics, Ethics, and Public Policy at the McDonough School of Business at Georgetown University, USA. He specializes in politics, philosophy, and economics and is the author of 16 books, including *Why It's OK to Want to Be Rich* (Routledge, 2020), *Markets Without Limits* (Routledge, 2016, with Peter Jaworski), and *Why Not Capitalism?* (Routledge, 2nd Edition, 2024).

Richard Yetter Chappell is Associate Professor of Philosophy at the University of Miami, USA. His primary research interests are in ethical theory, especially the defense and development of consequentialism. Chappell is the author of *Parfit's Ethics* (Cambridge University Press, 2021), and he blogs about moral philosophy at <goodthoughts.blog>.

Ryan W. Davis is Associate Professor of Political Science at Brigham Young University, USA. He writes about the value of autonomy in ethics, politics, and religion. He is the author of *Why It's OK to Own a Gun* (Routledge, 2024).

QUESTIONING BENEFICENCE

Four Philosophers on Effective Altruism and Doing Good

*Samuel Arnold, Jason F. Brennan,
Richard Yetter Chappell,
and Ryan W. Davis*

NEW YORK AND LONDON

Cover image: *Hand of a begging homeless man held out in appeal.* © RapidEye/Getty

First published 2025
by Routledge
605 Third Avenue, New York, NY 10158

and by Routledge
4 Park Square, Milton Park, Abingdon, Oxon OX14 4RN

Routledge is an imprint of the Taylor & Francis Group, an informa business

© 2025 Samuel Arnold, Jason F. Brennan, Richard Yetter Chappell, Ryan W. Davis

The right of Samuel Arnold, Jason F. Brennan, Richard Yetter Chappell, and Ryan W. Davis to be identified as authors of this work has been asserted by them in accordance with sections 77 and 78 of the Copyright, Designs and Patents Act 1988.

All rights reserved. No part of this book may be reprinted or reproduced or utilised in any form or by any electronic, mechanical, or other means, now known or hereafter invented, including photocopying and recording, or in any information storage or retrieval system, without permission in writing from the publishers.

Trademark notice: Product or corporate names may be trademarks or registered trademarks, and are used only for identification and explanation without intent to infringe.

ISBN: 978-1-032-83159-6 (hbk)
ISBN: 978-1-032-83155-8 (pbk)
ISBN: 978-1-003-50806-9 (ebk)

DOI: 10.4324/9781003508069

Typeset in Sabon
by codeMantra

CONTENTS

Introduction: Questioning Beneficence 1
Samuel Arnold, Jason F. Brennan,
Richard Yetter Chappell, and Ryan W. Davis

PART I
The Political Perils of Doing Good 5
Samuel Arnold

1 The Political Perils of Doing Good 7
  Samuel Arnold

2 Brennan's Response to Arnold 27

3 Chappell's Response to Arnold 36

4 Davis's Response to Arnold 43

5 Arnold's Response to Brennan, Chappell, and Davis 49

PART II
Effective Altruism and Regular People   **59**
Jason F. Brennan

6 Effective Altruism and Regular People   61
   Jason F. Brennan

7 Arnold's Response to Brennan   82

8 Chappell's Response to Brennan   89

9 Davis's Response to Brennan   97

10 Brennan's Response to Arnold, Chappell, and Davis   103

PART III
Exploring Beneficence   **113**
Richard Yetter Chappell

11 Exploring Beneficence   115
   Richard Yetter Chappell

12 Arnold's Response to Chappell   129

13 Brennan's Response to Chappell   137

14 Davis's Response to Chappell   145

15 Chappell's Response to Arnold, Brennan, and Davis   151

PART IV
Limiting Beneficence 161
Ryan W. Davis

16 Limiting Beneficence 163
   Ryan W. Davis

17 Arnold's Response to Davis 179

18 Brennan's Response to Davis 187

19 Chappell's Response to Davis 197

20 Davis's Response to Arnold, Brennan, and Chappell 207

Bibliography 217
Index 229

INTRODUCTION

Questioning Beneficence

*Samuel Arnold, Jason F. Brennan,
Richard Yetter Chappell, and Ryan W. Davis*

Most of us live high while others die. Should we help them? If so, how? And how much?

Most of us also already *do* help others, whether through charity, volunteering, or our productive work. Are we doing enough? Is our giving, volunteering, or work *working*, or might we instead be making the world worse even though we intend to help?

In recent years, the philosophical and activist movement called Effective Altruism has reinvigorated the debate about the nature of beneficence. The movement has exposed that many forms of giving do little good, many do active harm, but others do tremendous good.

This book debates Effective Altruism and beneficence more generally. Must we be effective altruists, or is effective altruism a mistake? How much should we give, or should we act beneficently through our normal work or political behavior? What are the respective roles of different kinds of institutions?

This book is an open-minded debate among four philosophers with different views about beneficence and Effective Altruism. Samuel Arnold is worried about the possible dangers of the movement and of private philanthropy. Jason Brennan comes closest to having the commonsense or default view of beneficence, but nevertheless thinks most people should adopt some version of Effective Altruism. Richard Yetter Chappell has a more revisionary and demanding view of ethics; in his view, beneficence is the primary moral principle. Finally, Ryan Davis contends that while effective altruism can be praiseworthy and good, nevertheless, beneficence toward strangers is nearly always above and beyond the call of duty.

Books of debate are nothing new in philosophy. But, as of now, "debate"-style books in philosophy tend to take one of two formats. There are many books in which two authors stake out a position over three to four chapters and then write brief, one-chapter responses to each other. There are also many anthologies, which offer "debates" only in a very broad sense that authors write independent papers with little interaction.

This book instead offers a new format: *the roundtable* or *the forum*. Each of us takes the lead in producing and answering three questions on the topics of beneficence and Effective Altruism. In turn, each of the other three authors will write a response to those questions. Finally, the author of those three questions responds to all of the responses.

Twelve Main Questions

These are the 12 main questions we will debate in Parts I–IV:

From Samuel Arnold:

1. Can philanthropy be unjust?
2. Can philanthropy be undemocratic?
3. Should businesses support controversial socio-political causes?

From Jason Brennan:

4. Can we endorse Effective Altruism without endorsing utilitarianism?
5. Is beneficence confined to charity and volunteering, or can we exercise it in other domains, including business?
6. Why are people so bad at charity and what should we do about it?

From Richard Yetter Chappell:

7. How important is beneficence compared to other values?
8. Does self-sacrifice make beneficence more virtuous?
9. How much should we care about future generations?

From Ryan Davis:

10. Is there a moral obligation to beneficence?
11. Is Effective Altruism morally risky?
12. Can Effective Altruism be part of a meaningful moral life?

As readers can see, we are not trying to win. We are thinking out loud and trying to learn from each other. We are taking intellectual risks. As we go, we each admit our uncertainties and at times change our views in response to others' ideas. We respect each other, the field, and ourselves too much to insist on getting the last word.

A Note to the Readers

Throughout this book, we will capitalize "Effective Altruism" when referring to the explicit philosophical and activism movements, and we will capitalize "Effective Altruist" to refer to someone who explicitly subscribes to that philosophy or movement. We will leave the terms uncapitalized to refer more broadly to altruism that works or to people who do effective work.

One reason for doing so is that we don't want to beg any questions. "Effective Altruism" is a substantive view with real commitments; it's not simply the view that beneficence should be effective in some minimal sense. Critics of Effective Altruism often criticize it *because* they dispute whether its recommendations are actually *effective*. Accordingly, we want to leave open that someone could meaningfully say, "I am an effective altruist but not an Effective Altruist; I reject Effective Altruism because its ideas are ineffective." For instance, Sam Arnold, one of the four authors, is certainly not opposed to being effective in his altruistic endeavors, but as readers will shortly see, he has many substantive complaints about Effective Altruism, the movement.

PART I
The Political Perils of Doing Good

Samuel Arnold

1

THE POLITICAL PERILS OF DOING GOOD

Samuel Arnold

Donating to charity. Running a business with an eye to the public good. These seem like excellent ways to benefit others.

Often, they are. But beneficent action, even when effectively designed and targeted, may fit poorly with political ideals we rightly hold dear. One such ideal is justice. Another is democracy.

My entries explore the tension between these political ideals and beneficence. Overall, my thesis is that it's rather hard to do good at scale without *also* doing something a bit dodgy from political morality's perspective. Beneficent action potentially casts a dark shadow in the realm of political values. That doesn't, of course, mean any given act of beneficence is unjustified, all things considered; but it *does* mean that the question, "Should I perform this beneficent act?" is more complicated than one might have supposed.

I'll start with the surprisingly complex relationship between philanthropy and justice.

Question 1: Can Philanthropy Be Unjust?

Reich defines philanthropy as "the deployment of private assets towards a public purpose."[1] What I'll call the *Default View* of philanthropy grants donors "wide [moral] discretion in deciding how to direct charitable donations."[2] Presumably there are *some* moral limits here: you shouldn't donate money to, say, an ethically deranged terrorist group. But within broad boundaries, donors may send their money wherever they want, for whatever reasons they please.

The Default View is incorrect. Donors do *not* enjoy wide moral discretion to give as they please; instead, their giving is bounded by at least two important moral constraints. They should, as Effective Altruists have insightfully argued, give *well*; they should strive to "do the most good with whatever resources [they] have."[3] But more than this, they should give *justly*.

Or so I will argue below.

Beneficence Constrains Donor Discretion

Suppose you're fixing to donate. According to the Default View, good on you: donating, no matter its target, is always laudable, and you're morally free to direct your funds however you see fit. Perhaps, like many Americans, you'll donate to your church, school, or symphony. Whatever you choose, no worries. It's your money; you're morally entitled to donate it as you please.

This ethical leeway is completely unsustainable, I think, in the wake of work by Peter Singer and other Effective Altruists.[4] Singer's famous drowning child argument shows that it's wrong to spend on frivolous personal consumption. That $500 you spend annually on lattes could, over a decade, save a life. Is 10 years of luxury coffee morally equivalent to a stranger's life? Of course not. The moral opportunity cost of frivolous personal consumption is, Singer shows, simply too high.

But the same logic condemns inefficient philanthropy. By giving up lattes, you can save a life. By redirecting badly-directed charity, you can save—well, potentially *hundreds* of lives, if you're really rich. Suppose you were planning to donate $50,000 to help renovate your *alma mater*'s football facilities. While this isn't the worst use of money, it's far from the morally best. Send that check to GiveWell, and thousands of kids won't get malaria.

So what's more important, morally: a shiny new locker room, or saving lives and reducing disease? The answer is obvious.

The Default View, then, is wrong; as Effective Altruists have shown, donors are *not* morally free to give however they like.

Instead, a norm of efficient beneficence constrains donor discretion. As Larry Temkin puts this point, "In giving to charity, one *should aim at doing as much ... good ... as one can*, given one's resources."[5] That italicized "should" clause, note, packs an ethical punch. Inefficient giving isn't merely wasteful; it's morally wrong.

Justice Constrains Donor Discretion

Suppose you're about to donate to GiveWell, which—let's stipulate—is the most effective charity on offer. This sounds morally impeccable. But there's a twist. *You just robbed a bank*; that's where you got the money. Are you morally entitled to donate it? Clearly not! It's not yours. You shouldn't give it away, even maximally well; instead, you should return it, thereby repairing the injustice you've caused.

This simple example shows something important: beneficence is not the only value constraining donor discretion. Justice matters, too. And, indeed, it seems—in some cases, at least—to matter *more* than beneficence. In the robbery case, welfare is probably maximized by giving the ill-gotten funds to GiveWell. But that would serve justice poorly, and it seems that justice is, in this case anyway, paramount. It's morally better to return the money, leaving some welfare gains on the table.

If all this reasoning showed is that stolen money can't be donated, well, it wouldn't be very important. But the reasoning extends further. Consider a case discussed by Temkin:

> *Broken Window:* A group plays baseball in a vacant lot. A hit ball shatters a nearby car's window, causing $500 in damage.[6]

These players are morally obligated to repair the damage. Failure would be unjust. This deontic requirement, notice, constrains the players' spending. There's a kind of *deontic moral lien* on their money. Having wronged another, they're in moral arrears. Paying off that moral debt takes priority over other projects—even beneficent ones. Were the players to send the $500 to GiveWell rather than fixing the car, that would be wrong.

As Cordelli puts this point, "debtors cannot be choosers."[7] Only *after* discharging their deontic duty of reparation do they regain moral discretion over their resources. We can summarize this line of argument in a principle:

> *Discretion only if no lien*: Donors enjoy moral discretion over their resources only if there is no deontic moral lien on those resources.

"OK," some readers may be thinking. "Thieves and window-smashers should settle their moral accounts prior to donating. But that doesn't apply to *me*—or, presumably, to many other would-be philanthropists."

This response fails to appreciate how broadly the above principle applies. I suspect that every one of us has *multiple* "deontic moral liens" on our resources—even those of us who haven't stolen, or smashed a window with an errant baseball.

For consider: the deep reason why there's a moral lien on resources in the stolen money case, or in the smashed window case, is that those cases (1) involve agents who are *complicit in injustice*; and (2) the *repair of which requires resources*.

But upon reflection, that description fits pretty much everyone reading this chapter.

Everyday Complicity

Here's the argument.

1. "Citizens of legitimate, democratic states are complicit in injustices committed by their states."[8]
2. The United States—a legitimate, democratic state—has committed, and is committing, serious injustices.[9]
3. Therefore, citizens of the United States are complicit in serious injustices.

Assuming that you're an American citizen, the argument establishes that you *are* complicit in injustice; hence, there's a moral lien on your resources, which constrains your donative discretion.

To defend premise 1, consider:

Catastrophically Bad Lifeguard. Some neighbors form a pool club, run democratically. The club deputizes a personnel manager, who hires Billy, a lifeguard. Billy, it turns out, is a catastrophically bad lifeguard. He not only fails to save several children from drowning, he actually pushes several others into the drink, actively killing them. Despite his misconduct, the personnel manager doesn't fire him; nor do the club's members fire the personnel manager. Everything continues as before, with Billy drawing checks and children drowning.

Clearly Billy is a moral monster, guilty of serious injustice. But notice that there's plenty of moral blame to go around. The personnel manager, surely, is complicit in Billy's crimes. *As are the club's members*, who, after all, deputized the personnel manager, and ultimately *bear moral responsibility for the pool club and its actions.*

Are they *as* bad as Billy? No. But they're clearly *more* complicit than some random neighbor who *didn't* join the pool club.

But if members of a democratic pool club are complicit in their club's injustices, surely members of a *democratic state* are similarly complicit in their *state's* injustices. Hence, premise 1.

Premise 2 receives support from across the ideological spectrum. If there's one thing political philosophers of all stripes can agree on, it's that the United States (like all countries) has a lot of unrepaired wrongs to answer for.

Consider some of the major theories of *domestic* justice:

- Libertarians, liberals, and radicals alike will condemn the obvious, unrepaired *historical* injustices like slavery and the conquest of Native Americans.
- Liberals would point to ongoing *distributive* injustices stemming from the country's failure to eliminate poverty and its highly unequal distribution of opportunity, affluence, and power.[10] Even libertarians like Nozick would blanch at the actually existing distribution of income and wealth—not because it's unequal, but because it didn't arise through just steps.[11]

- Theorists across the spectrum would highlight the country's morally disastrous "war on drugs," which has unjustly ruined millions of lives.[12]

Turning to *global* justice, we find more grist for premise 2's mill.

- Thomas Pogge argues that extreme global poverty, which results in tens of thousands of deaths daily, is the foreseeable and avoidable consequences of the global economic order imposed by rich countries (like the US) on poor countries. In other words, Pogge contends that each day, countries like the US commit grave human rights violation by, in effect, *condemning thousands to die* from poverty—poverty that rich counties *could easily eliminate* through relatively minor changes to the economic institutions they impose on developing nations.[13]
- "Open-borders" advocates like Carens, Huemer, and Hidalgo argue that immigration restrictions are unjust. If they're right, then the US—which restricts immigration—is guilty of serious, ongoing injustice.[14]
- "Just War Theory" lays down demanding conditions that wars must meet to be morally justified. The US has been involved in many wars that do not satisfy these conditions. These "unjust wars" have resulted in unfathomable amounts of unrepaired wrongdoing.

I trust the reader gets my point: the US—again, like all countries—has committed and is committing many injustices, just as premise 2 says.

But if premise 1 and 2 are true, then so is the argument's conclusion: Americans are complicit in serious injustice.

In other words, a state acting in our name, and under our democratic control, has committed grave injustices. We ordinary Americans are on the hook for these wrongs. Are we *as* complicit as powerful politicians in charge? No, but neither are we as innocent as, say, citizens of Fiji. We bear some responsibility. We have moral debts to pay. These debts must be borne in mind when spending one's resources—whether on personal consumption or on philanthropy.

Bottom line, *if you're an ordinary American, there's a deontic moral lien on your money*. Only after clearing your moral books will you be restored to a baseline level of moral discretion over your resources.

Ethical Philanthropy: No Easy Thing

According to the Default View, donors enjoy broad donative discretion. The Default View is wrong. First, as Singer and others have shown, donors must give *well*. Second, as Cordelli and Temkin have emphasized, donors must also give *justly*, where this means tending to any outstanding "deontic moral liens" on their resources, any unpaid moral debts that require attention *prior* to giving.

Fully moral philanthropy is, it seems, a rather complex affair. Donors must balance not merely demands of beneficence but also of justice. They must strive not merely to improve the lot of humanity—and what of non-human animals? Or future people?—but also to repay any past-due moral bills. What would this look like in practice? A worthy question—one that, perhaps, my co-authors and I can address in what follows.

Question 2: Can Philanthropy Be Undemocratic?

My first entry introduced the *Default View* of philanthropy, according to which, donors have wide moral discretion over their giving. I argued against this view, identifying two moral constraints on giving. First, as Effective Altruists have emphasized, donors should give *well*. Second, donors should give *justly*, where this may involve repaying outstanding moral debts prior to philanthropizing.

Here, I'll add a third constraint to the list. Donors should give *democratically*. In particular, they should strive to give without undermining the core democratic norm of political equality.

Elite Philanthropy

In philanthropy as in other areas, the rich are different from you and me. What we might call *ordinary philanthropy* involves relatively small donations, given unconditionally. Your aunt tosses spare change into her church's collection plate: that's ordinary philanthropy. Very different is what I'll call *elite philanthropy*, which involves massive donations, typically given with strings attached—as when Mark Zuckerberg gave $100 million to Newark schools, "on condition that the district embrace a slate of market-friendly reform proposals."[15]

Elite philanthropists donate truly staggering sums to a wide array of causes. Mike Bloomberg, for instance, gave away some $462 million in 2014, much of it going to fight climate change and improve public health.[16] In my own neck of the woods, the billionaire Bass family routinely donates millions to support cultural institutions like Texas Ballet Theater. Much elite philanthropy is channeled through private foundations established and controlled by extremely rich individuals, such as the influential Gates Foundation, which has disbursed tens of billions since 2000.

Massive gifts can do a lot of good. But they also tend to provoke an inchoate sense of unease. Take Zuckerberg's Newark gift. It's undeniably generous, but isn't there something vaguely unsettling about it?

Many philosophers have thought so. Billionaire philanthropy, some argue, merely treats the symptoms of a deeper disease.[17] Plugging gaps in school funding will never solve the root problem, which is background economic injustice.

Others object to the way elite philanthropy lets ordinary citizens off the moral hook.[18] Some obligations are meant to be discharged by specific agents. *I* have a duty to care for my son; if you swoop in and do it for me, my son is cared for, but not—one wants to say—by the right person. Similarly, an area's schools are supposed to be funded by *residents*, not some out-of-town billionaire. Something important is lost when Zuckerberg discharges what should be a communal obligation.

While these arguments help explain the intuitive unease many people feel when learning of the latest eight- or nine-figure philanthropic intervention, they don't get at what I take to be the deepest problem with elite philanthropy: namely, its undemocratic nature.

Elite Philanthropy Is Plutocratic

To see the tension between elite philanthropy and democracy, put yourself in the shoes of someone with kids in Newark schools. From experience, you're familiar with the system's strengths and weaknesses. You may even have ideas about possible reforms. But at the end of the day, you're just one ordinary citizen among many; your ideas—good, bad, or indifferent—count the same as anyone else's.

Or do they? Some billionaire comes along with a massive gift, and remakes the system in his preferred image, virtually overnight. *His* ideas, it turns out, matter vastly more than yours. Why? Not, at root, because they're better or more popular, but because they come attached to a gigantic pile of money.

Everyone grumbles about the state of the world, but most people can't really do anything about it. Not so with the ultra-rich. Their ability to (as David Callahan puts it) "give on a scale, and with an impact, that most people can't imagine" endows them with what sociologist Paul Schervish calls "hyperagency": the capacity to actually change the world.[19] "We love the fact that we're able to do something," reports billionaire philanthropist Laura Arnold. "Instead of just being outraged, we try to channel that outrage into a solution that we think is constructive and a better alternative."[20]

Such power! It's as if the elite philanthropist has a magic wand for conjuring social change out of thin air. "What takes a social movement for ordinary people to accomplish," Schervish notes, "wealth-holders can do relatively single-handedly."[21] A billionaire can hear about a social problem one day and solve it the next. Meanwhile, ordinary citizens can—what? Donate a few dollars? Volunteer? Call their Congressperson?

At this point, the democrat's alarm bells are blaring: *disturbing political inequality detected!* Notice that the democratic complaint here *isn't* that billionaires are bad at solving social problems, or that billionaire philanthropy is substantively misguided. No, the democratic objection is that such philanthropy is *procedurally unfair*. It's not right, the democrat insists, for a tiny elite—no matter how politically competent and wise—to enjoy vastly disproportionate influence over issues that affect all of us.

Not, at any rate, in a democratic society where we purport to be civic equals. In such a society, "citizens should be able to participate as equals in deciding how society will address important issues of common concern."[22] When deciding Newark school policies, for instance, Mark Zuckerberg's opinion shouldn't count for vastly more than an ordinary resident's.

We have a word for a society where—as Saunders-Hastings describes—"outcomes of common concern … are overwhelmingly responsive to the preferences of the rich": *plutocracy*.[23] At root, the democratic objection to elite philanthropy is that it threatens to transform what should be a *democratic* society—one where, "as equals, members … share authority over their common life and matters of common concern"[24]—into a *plutocratic* one, where "hyperagential" millionaires and billionaires enjoy an almost magical ability to shape the world to their liking.

The Argument, Formalized

To formalize the argument:

1. On pain of undermining political equality, issues on the democratic agenda should be decided through a broadly egalitarian process, i.e., one that grants everyone a roughly equal say.
2. When elite philanthropy targets issues on the democratic agenda, it renders the decision-making process concerning those issues inegalitarian.
3. Therefore, when elite philanthropy targets issues on the democratic agenda, it undermines political equality.

Notice that, per premise 1, not *all* issues need to be decided through an egalitarian process—only those "on the democratic agenda." But which are these? According to the mainstream view in democratic theory, *any issue involving important matters of common concern* falls on the democratic agenda, and should therefore be decided through a broadly egalitarian process. Such issues include, surely, not *just* electoral decisions (like which candidate to send to Washington) but also decisions about economic policy, foreign affairs, infrastructure, social justice, education, health care, and countless other intuitively political topics.

Notice that elite philanthropy tackles all these issues. For instance, Zuckerberg weighs in on education policy; the Gates Foundation shapes global health care; and the Bezos Day One fund pursues affordable housing. Hence, the first clause of premise 2 is sometimes satisfied. Elite philanthropy *does* sometimes target issues on the democratic agenda.

But according to premise 1, such issues are to be resolved through a broadly egalitarian process. Elite philanthropy, per premise 2, mucks that up. It renders what should be an egalitarian process anything but.

Evaluating the Argument

Premise 2 seems unassailable. How can one deny that, when billionaires weigh in with conditional giving on a topic, they enjoy vastly disproportionate influence over that topic? Just think of Zuckerberg's influence over Newark schools.

To resist the argument, then, one must reject premise 1. But that's an unpromising route. After all, premise 1 simply points out that democracy requires some degree of political equality. It says that, if we want to live in a democratic society, we need to give people an equal say over—or, as some philosophers put it, an equal opportunity to influence—important issues of common concern. I take this idea to be extremely well subscribed among democratic theorists.[25]

Exceptions, Qualifications, and Unresolved Questions

Drawing on work by Saunders-Hastings, Lechterman, Callahan, and others, I've argued that elite philanthropy—the giving of massive, conditional gifts—threatens political equality, a key democratic norm. As Callahan writes, "giving by the wealthy is amplifying their voice at the expense of ordinary citizens, complementing other tools of upper-class dominance" (287).

However, not *all* elite philanthropy undermines political equality. Consider philanthropy that targets intuitively private issues, issues not on the democratic agenda: donating to a symphony, say. There's nothing undemocratic about that. Democracy doesn't require that everyone has an equal say over *everything*—only an equal say over important issues of common concern.

Or again, consider unconditional gifts that merely support existing, democratically-chosen initiatives. For example, in 2019, Ray Diallo and his wife donated $100 million to the state of Connecticut to plug gaps in education and anti-poverty funding while "leaving existing policies on these matters untouched."[26] Through such gifts, billionaires don't drown out the voices of ordinary citizens; instead, they better enable ordinary citizens to accomplish the goals they've democratically selected. Granted, we might worry that, over time, any polity dependent on billionaire philanthropy—even in this hands-off, "the people pick, the billionaire funds" form—will veer in a plutocratic direction. But viewed as a singular event, a Diallo-style gift seems compatible with political equality.

So not *all* elite philanthropy offends against democratic norms. But much of it does. Some readers may be inclined to retort: so what? How important is political equality, anyway? My (necessarily brief) answer is that it's one important value among others. We shouldn't forfeit it lightly; but neither should we secure it no matter the cost.

Ideally, we'd keep the baby of billionaire beneficence while tossing out the bathwater of plutocracy. We'd find a way to preserve democracy while also giving rich people ample scope to use their fortunes for good. Various reforms have been suggested; perhaps we'll explore them further in the ensuing discussion.

Question 3: Should Businesses Support Controversial Socio-Political Causes?

"Corporations," remarks a recent study, "have changed."[27] Previously allergic to political controversy, corporations are newly determined "to change the world, one pressing issue at a time." To this end, they rush headlong into "corporate activism," defined as "a firm's public demonstration of support for or opposition to one side of a partisan socio-political issue."[28] Consider a few examples:

- In a famous ad, Gillette channels #MeToo, decrying "toxic masculinity" and admonishing men to do better.
- During Pride Month, Target releases special LGBTQIA+ merchandise, including kids' clothes and "tuck-friendly" women's bathing suits.
- Major League Baseball boycotts Georgia over a controversial election law.

In these and many other instances, corporations have proven surprisingly eager to lob grenades in our raging culture war.

What should we make of this important new trend in corporate behavior? Broadly speaking, I'm not a fan. Against corporate activism, I'll raise two complaints: first, it's undemocratic; second, it adds to the undesirable over-politicization of everyday life—what Talisse calls the "political saturation of social space".[29]

Business's Social Purpose

Question: What's the purpose of a business? Standard answer: Businesses exist to make money for their owners.

Proponents of "stakeholder capitalism" would give that answer only partial credit. On their view, companies must do more than merely turn a profit. They should also "serve a social purpose [by benefiting] all of their stakeholders, including shareholders, employees, and the communities in which they operate."[30]

What kinds of social benefits do stakeholder capitalists have in mind? Recent scholarship distinguishes two models of corporate beneficence. "Corporate Social Responsibility" asks businesses to further the common good by producing "universally recognized benefits for all stakeholders."[31] These are things like cancer awareness drives—a-political, uncontroversial goods or services whose value even partisan foes can acknowledge. Companies can provide such goods without thereby taking sides in the culture war.

But perhaps they *should* take sides in the culture war? That's the core idea behind a second model of corporate morality, "corporate activism," which urges companies to pursue even politically divisive social improvements. By all means, corporate activism says, fund the food bank—but also stand ready to save the planet, pursue social justice, and defend democracy. "We know what needs to be done to build a just and sustainable world," writes Rebecca Henderson, one of corporate activism's preeminent theoreticians. "Business must step up."[32]

Curiously, corporate interventions on these divisive issues manage to be astonishingly one-sided. No matter the topic—abortion rights, racial justice, LGBTQ issues, etc.—corporate America "invariably throws its considerable weight behind cultural progressives," as Darel Paul notes.[33] We live in an era, not merely of corporate activism, but *ideologically monolithic* corporate activism: what Ross Douthat calls "woke capitalism," a historically unprecedented politico-economic alliance between global mega-corporations and progressive elites.[34]

Paths Not Taken

Here, I'll focus on corporate activism, leaving consideration of its less divisive cousin, corporate social responsibility, for another time.

One way to evaluate corporate activism is to do so on a case-by-case basis, approving of substantively correct activism while disapproving of substantively incorrect activism. That's not my approach. I find fault with *all* corporate activism, no matter its ideological valence or substantive merits.

In the next two sections, I'll build the democratic case against corporate activism. Then I'll unpack the "overpoliticization" objection.

Corporate Activism Is Undemocratic

Recall the structure of my democratic complaint against billionaire philanthropy.

I began with a key democratic principle, namely, that issues on the democratic agenda must be decided on a broadly egalitarian basis. Otherwise, we don't have political equality, a crucial element of a democratic society.

I then noted that elite philanthropy violates this principle. Elite philanthropists use their economic power to exert disproportionate, political-equality-deranging levels of influence over issues of common concern. Think of Zuckerberg determining Newark school policy, thanks to his conditional gift. That's undemocratic.

As is corporate activism, and for the same basic reason: namely, it undermines political equality—as I'll now explain.

As a first step, let me further unpack this idea of political equality. Suppose a group aims to decide something in an egalitarian way: say, where to dine. How might this process, *qua egalitarian* process, misfire? How could it fail to be egalitarian?

There are two ways.[35] First, *inputs could be disproportionate*. Some members could have too much say. Maybe you get 2 votes to my 1. Second, *collectively chosen outputs could be subverted*. Having decided together, in an egalitarian way, to go to Joe's Pizza, you might be able to override this result. Maybe you own the car we need to get there. "On second thought," you can say, jangling your keys, "let's go somewhere else."

Corporate activism undermines egalitarian decision-making in both of these ways. First, it renders inputs disproportionate. The root problem: "corporate persons possess dramatically more resources for expressing their positions than natural persons do."[36] This isn't just about money. Companies have massive war chests, but also prominent platforms from which to broadcast their socio-political messages, as well as a captive employee audience required to listen. Given these factors, it's no surprise that the ordinary person's voice is the faintest whisper compared to Globocorp's booming corporate megaphone. So when Globocorp weighs in on the divisive topic *du jour*, its input utterly dwarfs the average citizen's—a clear violation of democratic norms.

But the story gets worse for political equality. Globocorp, it turns out, doesn't just have a megaphone; it's also got a veto button, which it can use to simply override disliked policy change.

Consider Indiana's ill-fated Religious Freedom Restoration Act of 2015, which passed its legislature comfortably one week, only to be gutted by that same body the next. What explains the sudden reversal? In a word, the threat of *capital flight*. Major corporations like Apple and Salesforce, regarding the bill as anti-gay, "threatened to diminish or withdraw their economic presence in the state" unless the legislature backed down—which it did.[37]

Thus do activist corporations have, not only a megaphone with which to dominate political discourse, but also a veto button, which they can use to cow ostensibly sovereign polities into compliance with their policy preferences.

It's hard to see how any of this is compatible with political equality.

Doesn't Corporate Activism Simply Mirror Public Opinion?

However, there's more to democracy than political equality. Imagine that powerful elites were somehow constrained to use their disproportionate influence in ways congenial to ordinary people. Isn't there something recognizably democratic—even if not politically egalitarian—about an arrangement in which elite choices reliably track public opinion?

Inspired by this thought, one might defend corporate activism's democratic credentials as follows:

> *The Defense:* Corporate activism is actually quite democratic. After all, why does Starbucks fly the Pride flag? Ultimately, for the same reason it introduces a new latte—namely, *to make more money*. Activist companies are just catering to consumer demand; they're just driving up business by saying what customers want to hear. Hence, *corporate activism is ultimately controlled by, and responsive to, public opinion*. Surely that helps burnish activism's democratic chops. How undemocratic can corporate activism be, really, if companies take their activist-marching-orders from ordinary people? Indeed, isn't there a sense in which corporate activism, on this picture, is positively *good* for democracy? It helps amplify ordinary people's views on contested social and political issues.

The Defense is interesting, but ultimately unsuccessful.

First, it does nothing to blunt my main objection, which is *not* that corporate activism fails to mirror public opinion, but that it undermines political equality by giving corporations disproportionate influence. *The Defense* does not challenge this point.

Second, *The Defense*'s central empirical conjecture is false—indeed, obviously so. Corporate activism, the argument reassures us, simply mirrors public opinion. Since public opinion is ideologically diverse, corporate activism must be, too. Right?

Well, no. One looks for conservatively-inflected corporate activism in vain; strangely, it basically doesn't exist. Outside of a few economically marginal counterexamples—a My Pillow here, a Black Rifle Coffee there—activist corporations pin their colors exclusively to the progressive mast.

We've got a politically divided country inundated with exclusively progressive corporate messaging. Corporate activism doesn't mirror public opinion; it mirrors *part* of public opinion, the progressive part, leaving the views of moderates and conservatives entirely out of the picture. In other words, corporate activism *distorts* public opinion more than it reflects it.

Thus, contrary to *The Defense*, there's something importantly *undemocratic* about corporate activism's relationship to public opinion. Through corporate activism, powerful economic elites, far from faithfully channeling the views of "ordinary people" taken as a bloc, signal-boost an ideologically uniform subset of these views. Multi-billion-dollar companies are manning the trenches for *one side only* in our fractious culture war.

It's hard to see anything democratic about *that*.

Overpoliticization

Consider another problem with corporate activism: it furthers "the political saturation of social space"—the pernicious overpoliticization of everyday life, the creeping incursion of partisan experiences, signifiers, and shibboleths into formerly (and blissfully) a-political spaces.[38]

In these politics-soaked times, nearly everything we do seems freighted with political meaning. It's not just that partisan foes listen to different music, wear different clothes, shop at different stores, and watch different programs; it's that "these [everyday] choices [have become] ways of *publicly expressing* [one's] political commitments."[39] By listening to country music, shopping at Whole Foods, watching the NFL, or engaging in any number of ordinary, everyday actions, one picks sides in the culture war—whether one wants to or not. "Politics," writes Talisse, "has become *everything that we do.*"[40]

This constant and inescapable political signaling reinforces tribal identities, encouraging each side to define itself in opposition to the other not only politically but in virtually every aspect of life. Each side comes, not merely to lose solidarity with the other, but actively to dislike the other. As politics swallows everything, political foes turn into personal enemies. For example, in 2020, Pew found that 47% of Republicans wouldn't date someone who pulled the lever for Clinton, while 71% of Democrats wouldn't date a Trump voter.[41]

A sorry state of affairs, this. I take political saturation to be deleterious to human flourishing as well as an obstacle to democracy (it's hard to deliberate with those you hate).

What does this have to do with corporate activism? As Talisse stresses, what's needed to heal our partisan divide is *less* politics, not more. We need a-political spaces, venues for "nonpolitical cooperative endeavors" within which we can interact not as political opponents but simply as ordinary people, doing whatever non-political thing it is we're doing together.[42] By cooperating together in contexts where politics is simply irrelevant, we can come to see one another, not as civic enemies, but civic friends, reasonable people who can reside together despite deep political differences.

Corporate political activism, notice, torpedoes this healing strategy. The more Globocorp blares through its activist megaphone, the more politically saturated our social space becomes. Even a decade ago, one couldn't shop or work at a Patagonia or a Hobby Lobby without signaling partisan affiliation; now, as more and more companies fly partisan colors, the space for politics-free, negative-partisanship-reducing economic cooperation has dwindled precipitously.

Corporations Should Generally Avoid Activism

Stakeholder capitalists implore companies to do good even as they try to make a buck. I've urged caution. When companies weigh in on socio-political issues, they inevitably deny ordinary people the equal say they deserve. That undermines political equality. Moreover, when companies weigh in *monolithically*—when they all sing from the same ideological hymnal—they distort deliberation by amplifying a mere subset of public opinion. That makes our political process less representative. In addition, corporate activism contributes to the undesirable overpoliticization of everyday life. None of this is good.

What do I propose instead? Companies should generally "forbear" from corporate activism.[43] For the sake of democracy and human flourishing, they should largely steer clear of divisive socio-political engagement.

My advice? Wall Street, stop emulating Harvard Yard; CEOs, stop acting like gender studies professors. Just stick to business, corporate America, and leave the culture war to the culture warriors.

Notes

1. Reich 2019, 7.
2. Cordelli 2020, 239.
3. MacAskill 2016, 12.
4. See, e.g., Singer 1972.
5. Temkin 2022, 33.
6. Ibid., 287.
7. Cordelli 2020, 238.
8. Ibid., 246.
9. Notice that this premise could refer to pretty much *any* state; all states have done, and are doing, unjust things.
10. See, e.g., Rawls 1999.
11. Nozick 1974.
12. See, e.g., Huemer 2004.
13. Pogge 2007.
14. Carens 2015; Huemer 2010; Hidalgo 2018.
15. Lechterman 2022, 26.
16. Callahan 2017, 15.
17. Syme 2019.
18. Beerbohm 2016, 215.

19 Callahan 2017, 42.
20 Ibid., 42.
21 Ibid.
22 Hussain 2012, 118.
23 Saunders-Hastings 2022b, 72.
24 Ibid.,14.
25 Some readers may reject premise 1 as too egalitarian. No problem. The argument can operate with a weaker variant of premise 1, one that requires not an *equal* say, but merely a *not-wholly-outmatched* one. See Saunders-Hastings 2022b.
26 Lechterman 2022, 26.
27 Masconale and Sepe 2022.
28 Bhagwhat et al. 2020, 1.
29 Talisse 2019, 74.
30 Fink 2018.
31 Masconale and Sepe 2022, 262.
32 Henderson 2020, 27.
33 Paul 2022.
34 Douthat 2018.
35 See Saunders-Hastings 2022a, 194.
36 Lechterman et al., 2024,
37 Deneen 2023, 55.
38 Talisse 2019, 74.
39 Ibid., 92.
40 Ibid., 90.
41 Pew Research 2020.
42 Talisse 2019, 162.
43 See Saunders-Hastings 2022a.

2
BRENNAN'S RESPONSE TO ARNOLD

Question 1: Can Philanthropy Be Unjust?

As Arnold notes, EA is not complacent about philanthropy. The typical EA primer, such as William MacAskill's *Doing Good Better*, complains that many philanthropies do no net good, while many others cause more harm than good. Many charities and NGOs redistribute resources from rich people to other rich people. Weaker versions of EA, like the one I defend later in this book, insist we avoid bad charities. Stronger or more demanding versions of EA hold that by default we should give to the most productive charities. As Theron Pummer argues, if you can either save six strangers or one, you should save the six rather than the one.[1]

Arnold argues that various considerations of justice constrain our philanthropic giving. I agree in the abstract, even if Arnold and I disagree on what exactly justice is and about particular cases.

Indeed, I think there are yet other considerations besides justice and beneficence which rightly bear on our philanthropic giving and on our actions in general. I donate time and money to Woodson and Oakton High School's theatre groups, for instance, by chaperoning a four-day field trip for the former or performing guitar for the latter's musicals. By doing so, I redistribute time and money from me (a rich person) to benefit rich kids and rich audiences. I could have instead volunteered for needier programs. I could have given a few extra talks and then donated tens of thousands of honoraria dollars to GiveWell charities. But my goal

28 The Political Perils of Doing Good

here was not to promote justice or to act from pure beneficence, but to promote community and art. Beneficence helps us live; community gives us a reason to bother living.

I also read to my kids, buy them expensive presents, take them on expensive vacations, pay for expensive voice lessons, and so on. By doing so, I once again redistribute my resources to my privileged children. I could have instead provided for much needier children in West Virginia or DC, or donated more money to GiveWell charities. The goal here is not promote global justice or act from pure beneficence, but instead to be a good father and give my kids the best shot at a good life. Beneficence helps us live; family and fraternity give us reason to bother living.

Effective Altruists argue that many common philanthropic actions look lousy from the standpoint of pure beneficence. I agree, and I even agree that this is often reason to stop donating to one charity and instead pick another. I also think people tend to have selfish motives when engaged in philanthropy.[2] I am not complacent about people's philanthropic behaviors.

But I also acknowledge that many times people engaged in philanthropy are motivated by, and succeed in fulfilling, moral principles of reciprocity, community, fraternity, and the like, rather than principles of beneficence or justice. It's an interesting theoretical question whether pure beneficence or considerations of justice should always trump these other concerns, though I know of no philosophers or activists who act like they always do.

Arnold worries that citizens of modern democratic countries, such as the US, are complicit in their governments' injustices. To his credit, Arnold does not claim, as so many political theorists mistakenly do, that mere inaction in the face of injustice is complicity. Arnold recognizes that to be complicit in another's wrongdoing, one must actively participate in that wrongdoing by counseling, encouraging, or aiding the wrongdoer in specific ways. Not every action that causally contributes to a crime counts as complicity. (A taxi driver who unknowingly drives a murderer to the victim's house is not complicit.)

Instead, Arnold claims that we bear responsibility by participating in democratic politics. He compares citizens to members of a pool club who employ a horrible lifeguard.

Now, my first solo-authored book in 2011 argued at great length that citizens have a duty not to vote badly, and that citizens bear some blame for what their elected leaders do.[3] I've made a career out of bashing voters.[4] So you might expect me to side with Arnold. But I worry there are some disanalogies which undermine his argument.

First, being a democratic citizen is not like being a member of pool club, and the disanalogy is morally significant. People voluntarily join the pool and could leave at will. But most people around the world are stuck being citizens of whatever country they were born in; only a tiny minority have any real option to emigrate to another country.

Further, imagine all pool clubs have evil lifeguards, just as (I think Arnold agrees) all or nearly democratic and undemocratic countries commit serious injustices. We can simply refuse to join any pool club, but we are stuck living under some unjust government or other.

Second, Arnold has us imagine the pool club members do little or nothing to stop the injustice; indeed, they support the lifeguard. But in democratic countries, many people vote against injustice, join reform parties, try to vote for the lesser evil, engage in activism, and so on. They do the best they can, given the background constraints. Some don't vote or participate at all. These citizens are not complicit.

You might say that all citizens pay taxes which support injustice. But taxes, unlike pool club dues, are involuntary. I want to withhold my taxes from the DEA or the border patrol, but I can't. The US government threatens me with violence and removes income from my paycheck without my consent. By paying taxes, I am less complicit in its wrongdoing than a shop-owner is complicit with the Mafia after a shake-down.

Third, the pool club members have far more effective power than citizens do. In Arnold's thought experiment, it's obvious that the lifeguard is bad. It's easy to remove him and replace him with someone better. It's easy for pool club members to control the lifeguards. (I know all this firsthand as a voting member of a swim and tennis club.) Arnold thus needs to show us that fixing America's injustices is as easy as removing, replacing, or reforming a bad lifeguard, and that citizens have the kind of power pool club members have. Without that, his analogy fails.

Fourth, I'm not sure what the upshot is of Arnold's argument is. Suppose we grant that Americans are complicit in their government's injustice. Great. Does this mean they shouldn't donate to GiveWell charities? Does it mean they must use their money to fight injustice instead? (How?) How does this bear on EA?

Question 2: Can Philanthropy Be Undemocratic?

I almost feel sorry for billionaires. If they invest their wealth, people say they are hoarders. If they consume it, people say they live high while people die. If they engage in philanthropy, people complain that their giving is undemocratic and they exert too much power and influence. (The solution, they add, is to let *us* spend their money for them.) People who think billionaires are inherently bad conclude that whatever they do is bad.

It's analogous to how anti-market people complain about prices. If a company undercuts competitors, that's "predatory pricing." If a company prices its goods the same, that's collusion. If it prices things higher, than monopolistic. (The solution, they add, is to let us control what the companies do.)

Or, if I live in the suburbs, activists say that's white flight, but if I move to Anacostia, that's gentrification. (The solution, they add, is for me to feel guilty, give them money, and do what they say.) Damned no matter what you do.

Arnold's answer to his second question seems incompatible with his answer to the first. In response to question 1, Arnold argued at length that citizens in the US and other democratic countries are like idiot, negligent, and/or malicious pool club members who keep a murderous lifeguard in power, who manage a death trap, and who could easily do better but don't. Why, then, would he be worried rather than relieved that billionaire philanthropy is undemocratic?

To call something is democratic is not to say it's good or just; to call something is undemocratic is not to say it's bad or unjust. Socrates was condemned to death by a fair democratic assembly, but that action was evil. The US Bill of Rights, judicial review, and liberal limitations on the scope of government are each undemocratic, but these are generally good things. It would be more democratic to let elected committees plan the economy and set prices, but this would be disastrous.[5] Taylor Swift has massive influence over world culture—more than most democratic governments have—but few people think her cultural influence robs democratic governments of their rightful power. Certain things are beyond the rightful scope of democracy.

Arnold has various complaints about billionaire giving. Some are particular (e.g., Zuckerberg's donation to Newark schools) and others are principled (billionaires have outsized influence). But what's missing from Arnold's argument is the essential piece: we need evidence that democratic control would, in real-life, be superior to billionaire philanthropy. Without such evidence, we have no reason to endorse his conclusions or even care whether billionaire giving is undemocratic.

Consider again Zuckerberg's attempt to fix Newark's schools. If democracy is so great, why were these schools in need of fixing? The very fact he needed to make a donation is condemnation of democracy and proof of persistent democratic government failure. Combined US government spending on K-12 education is already nearly US$1 trillion. Washington, DC, spends about $25,000 per K-12 student but has poor outcomes. Lack of government money isn't the problem here.

If we somehow liquidated and confiscated all US billionaires' wealth, we could fund the US government for about eight months.[6] When Elon Musk wanted to buy Twitter, internet activists said he could end world hunger for US$6 billion. But democratic governments around the world have far more money than all the billionaires, and yet haven't ended world hunger, cured malaria, dewormed the world, and done the host of other things charities do. Why think—especially given the list of injustices Arnold gave us above—that democratic governments would do much good with slightly more money? If they are so great, they would have already solved these problems with the massive resources they already have. They haven't, so either they can't in principle or they are incompetent in practice.

In fact, democratic governance suffers from systematic incompetence. Democratic voting is a collective action problem. How *we* vote matters; how any *one of us* votes does not. It's akin to pollution: our pollution matters, but your pollution doesn't. Thus, citizens tend not only to be radically ignorant and deeply irrational in how they process political information, but do not even vote on the basis of policy. Voting is mostly about social signaling. Politicians in turn have few epistemic checks to motivate them to perform well. They are incentivized to think short term and to do what sounds good rather than is good. They are liberated from the consequences of bad decisions, since they externalize the costs and since voters do not do much retrospective voting.[7] We thus have little reason to think democracies would do better than billionaire philanthropy with the billionaires' money. Perhaps this is why the significant literature on global aid tends to conclude that government-funded aid to the global poor backfires.[8]

Democratic governments tend to look inward, but billionaires are free to fund people with real need. As Jessica Flanigan and Christopher Freiman say, we can "think of the US government as a charity that helps provide retirement income and healthcare for people who tend to be richer than 90% of the world."[9]

Further, billionaire philanthropy enables large-scale projects that democracies would not and perhaps should not support. For instance, many religious, scientific, or artistic projects are worth completing, but should or would not get funding from governments. It's not their job.

Question 3: Should Businesses Support Controversial Socio-Political Causes?

I agree with Arnold. Even if businesses have the right to take political stances, there are many reasons why businesses should not, except perhaps over political issues that directly affect them (such as proposed regulations). These include:

1. It's good to have respites from politics. Since business is pervasive, making business political would make politics pervasive.

2. Businesses have no particular wisdom in tracking the truth about political matters. They don't *add* something valuable by participating.
3. Corporations might exert undue influence on their employees.

Let's consider each in turn.

Point 1

In politics, most of us are tribalistic and nasty. We form in-groups and out-groups and despise the other side. Politics is not about policy; it's about social status, forming in-groups and out-groups, jockeying for position, gossiping, slandering enemies, surveilling friends, signaling fidelity, and other repugnant behaviors. It's a realm where people are rewarded for bad behavior.[10] Politics divides us. So, all things being equal, we should want to have less politics, not more. We should try to keep politics as contained and restrained as possible.

Point 2

Consider a different group that meddles in politics—universities. It's now commonplace for university presidents to issue public statements about current events, such as the October 7, 2023, Hamas massacre of Israeli citizens. These issues are important and deserve serious attention. But it's not obvious university presidents should say anything about them.

For one, most university presidents have zero expertise on these matters. Many are career bureaucrats rather than scholars or experts. Their pronouncements are often vapid and platitudinous, offering neither insight nor solutions. Second, even when presidents are more competent, it's not their job to be pundits or social commentators.

Back to business: Corporate leaders are selected for their skills in making money, closing deals, managing production, marketing to consumers, and the like. They are not selected for their special insight into politics or morality. If you are unsure what justice requires, you wouldn't think to ask marketing executives at Adidas.

If businesses were embedded inside an incentive structure that made them especially likely to uncover important truths, that would be different. Consider, in contrast, that this is the main economic argument for having goods and services supplied by for-profit businesses on competitive markets. On the standard economic analysis, markets have *epistemic value*; businesses are in position to discover how best to satisfy people's disparate and changing desires. But for most socio-political causes, businesses have few advantages and many disadvantages.

Point 3

In the 1800s and early 1900s, the Left used to complain that bosses might force their workers to vote for corporate interests rather than labor interests. This was one classic argument for the secret ballot: Workers (and others) need to protect themselves.

So, there's a general argument here for corporate quietism about politics. It's often bad for corporate leaders, owners, managers, and the like to have outsized power and influence over politics. Realistically—especially given existing studies about political discrimination[11]—firms are likely to discriminate on the basis of politics when hiring. There are degrees of freedom here; I don't mean that the line supervisors will only hire workers who share the CEO's politics. Nevertheless, workers in a corporate environment face undue and corrupting political influence to keep their jobs. When corporations take political stances, this results in leaders pushing employees around in politics.

The corporate environment is unlikely to foster good political debate. Proper political deliberation requires people to meet each other as equals, free of outside and irrelevant forms of influence. This means no coercion and no threats. But the corporate environment is money-based, and often unequal and hierarchical. It's not a good environment for politics.

Notes

1 Pummer 2016.
2 See Simler and Hanson 2018.

3 Brennan, 2009; 2011.
4 Brennan 2016.
5 Hayek 1945.
6 Politifact 2021.
7 Achen and Bartels 2016; Brennan 2016; Caplan 2007; Kinder and Kalmoe 2017; Mason 2017; 2018.
8 Coyne 2013; Deaton 2013.
9 Flanigan and Freiman 2022, 763.
10 Simler and Hanson 2018; Mason 2018.
11 Iyengar and Westwood 2015.

3
CHAPPELL'S RESPONSE TO ARNOLD

Question 1: Can Philanthropy Be Unjust?

Arnold argues that reparative justice takes priority over general beneficence: you can't use stolen funds, or money owed to others, to promote the general good. He further argues that we are *all* complicit in serious injustices, and thus owe reparation prior to engaging in philanthropy.

I'm generally wary of moving uncritically between small-scale and large-scale rules. The reason for this is that the contexts differ sufficiently that they might well call for different rules. Consider the libertarian slogan: "Taxation is theft!" Theft is bad: laws against theft make for a more stable and successful social order. Is the same true of taxation? It seems not. There are some intrinsic similarities between the two, as libertarians emphasize. But the crucial question, at least for consequentialists, is whether the two have relevantly similar *effects*. There's nothing incoherent about thinking that a rule against theft would be welfare-promoting, but a rule against taxation would not.

Similar thoughts may apply to justice. It is plausibly welfare-promoting to require individuals to prioritize clearly-specified reparative duties over other interests and concerns in their everyday lives. But it doesn't follow that it would, for example, be welfare-promoting to expropriate all US land from its current owners to return it to Native American tribes. However "just" the latter upheaval might be, it would be insanely harmful. Similar remarks apply to the demand that poor African descendents of slavers pay reparations to wealthy African-American descendents of slaves. On a large scale, justice must always be tempered by beneficence.[1]

Moreover, I think it's very questionable whether we are in fact all complicit in injustices committed by our governments.

Arnold's defense of this claim depends upon an analogy to a democratically-run neighborhood pool club who hire Billy, a *catastrophically bad lifeguard*, who drowns many children: "Despite his misconduct, the personnel manager doesn't fire him; nor do the club's members fire the personnel manager. Everything continues as before, with Billy drawing checks and children drowning."

But this case is under-described. Suppose that the pool club is split into different factions. Some factions vociferously oppose Billy, but simply lack the votes to win out at the end of the day. Why should they be considered in any way complicit in a decision that they've opposed to the best of their ability?

Next, consider members of the pro-Billy faction. You might think that *they*, at least, must be complicit. But even this verdict is too hasty. Suppose that they know that the only other available lifeguard would be *even worse*: she would not only kill more children than Billy, but she would cover her tracks so well that they could never prove wrongdoing, and the other pool club members would insist on keeping her on, despite the ever-rising death count. If Billy is the lesser of two evils, and there really is no feasible alternative (it's well known that a majority would never agree to shut down the pool club entirely, for example), then keeping Billy in place might be the right thing to do after all. And anyone who does the right thing for the right reason is surely not "complicit" or blameworthy for the lesser harms that result, which they could not possibly have avoided without imposing even greater harms on others.

And, of course, this is precisely how many of us stand in relation to the injustices committed by our governments. Many injustices are committed by parties that we actively oppose. Others are committed by parties that we support, because the alternative party would (we believe) do even worse. Those who actively support the injustices in question may be complicit in them. As may those who negligently pass up an opportunity to successfully prevent the injustices (without expectably causing worse). But neither being a member of a democratic state, nor even voting for the party in power, entails support for the unjust things that the ruling powers end up doing. Nor do most citizens have any opportunity to successfully prevent their state's injustices (without expectably causing worse). So I don't see any good reason here to attribute such universal complicity.

Question 2: Can Philanthropy Be Undemocratic?

What should we think about donations with (policy-influencing) strings attached?

The answer may depend upon the generality of the question. If we expect most such strings to be distorting, biasing policy toward the interests of the donor rather than the general good, then we could reasonably oppose such strings *in general*.

But we can also evaluate particular proposals on a case-by-case basis. (Compare: some people are generally in favor of greater government regulation, others are generally against. But we might instead think the best approach is to focus on more specific questions, and evaluate any given proposed regulation on its merits: favoring good regulations and opposing bad ones. For an even starker example, consider how silly it would be to be blanket for or against *legislation*, without regard for the particular content of the legislation.) Maybe we should be opposed to strings that pull people toward bad policies, and favor strings that help to pull them toward better policies.

Arnold disagrees. He suggests that it is *procedurally unfair* for "a tiny elite—no matter how politically competent and wise—to enjoy vastly disproportionate influence over issues that affect all of us."

I'm not sure how much we should care about procedural unfairness. Suppose a benevolent and omniscient God offered to take over the reins of government for us. Would that be *unfair*? I'm not even sure what the question is asking. It would obviously be an improvement (only the most rabid partisan could possibly think that their preferred leader was *better than God*; and we tend to be landed with our dispreferred option half the time), and that's surely what really matters.

What's more, it just seems completely *inevitable* that some people will have more influence than others. Consider Tucker Carlson, Donald Trump, or members of the *New York Times* editorial board. No ordinary citizen has anything like the political influence of these culturally influential people. Is that "unfair"? Again, I don't know that the question even makes sense. People choose how to direct their own attention. Should we *force* them to put down their newspapers and listen to a homeless person instead? Surely not. Would it be "fairer"? Maybe, in some sense, but not in any sense that's worth caring about.

Given that strict equality of influence is a non-starter, I don't really see much force to the complaint that it's *unfair* that rich people (in addition to journalists, politicians, and other culturally influential people) have more influence than ordinary people. Fairness just doesn't seem the normative property that's worth attending to here. What we should assess is whether it's *harmful* that rich people have such influence, and if so, what should be done about it.

Even if it is *in general* harmful for rich people to have such influence, it doesn't follow that each *particular* rich person should refrain from trying to improve public policy. Perhaps the best way to stop a bad guy with a suitcase full of money is with a good guy's suitcase full of money.

Still, it could be worth thinking about ways to reconcile philanthropy and democracy. Usually, this involves trying to make philanthropy more democratic: Arnold approvingly mentions unconditional gifts to the state that leave existing policies untouched. But I'm skeptical of the wisdom of ordinary political priorities. I wonder if we might do better to make democracy more philanthropic.

Imagine decentralizing the public purse by distributing *philanthropic vouchers* to each citizen, to be regranted at their discretion to any eligible charitable organization or government-run project.[2] Like basic income proposals, philanthropic vouchers are a means of redistributing wealth and influence without necessarily increasing government. I think they have two primary benefits worth highlighting:

First, by making tradeoffs explicit, and making the stakes of one's decision explicit, it encourages more careful reflection and decision-making than typical political processes such as voting. Currently, the US government spends vastly more on the elderly than it does on children, and close to nothing on the global poor. It seems plausible to imagine that these priorities could change significantly if individuals had to explicitly decide for themselves how to apportion their share of the public purse. Or they might not: perhaps people would endorse the status quo distributions upon further reflection. But either way, such a process would surely have greater democratic legitimacy than our current political sausage-factory, in which most people have very little idea of what they are actually voting for when they elect one representative over another (except, perhaps, as it relates to the most hot-button issues).

Second, the incentives surrounding public political debate would be radically changed by shifting from winner-takes-all political contests to civic persuasion on a continuous scale, since now *each additional person you persuade makes an equal incremental difference to the overall distribution of resources.* Boosting support from 19% to 21%, or from 79% to 81%, is just as significant as the boost from 49% to 51% of the population. This is clearly more principled, and provides a salutary incentive to appeal to as wide an audience as possible rather than separating into competing political coalitions and resting content with 51% support.

Not all decisions can be decentralized in this way, so I don't pretend that this is a total replacement for all politics. And it raises a number of important questions, such as (1) what things should the government directly fund *independently* of citizen-supported funding? and (2) How restrictive should the eligibility criteria be for charities?[3] To address coordination problems, citizens might be invited to delegate their voucher to a trusted intermediary regranting organization (like the "EA Funds"[4] for effective altruists). Tracking running totals with live updates could better inform citizens about which of their priorities have already received sufficient funding from others' vouchers, and protect against wasteful overconcentration of funding on the most popular few organizations.

Many such details would need to be carefully worked out to make the idea perform as well as possible. But it strikes me as well worth exploring. Philanthropizing democracy would be a truly radical systemic change; but it's one with an unusual amount of promise. (And like basic income proposals, it could begin small, and gradually scale up if it seems to be working well.)

Question 3: Should Businesses Support Controversial Socio-Political Causes?

As indicated in my previous responses, I'm not much moved by the complaint that corporate activism is "undemocratic." I think we should be pragmatists, and support democratic norms and institutions in just those contexts where they can be expected to bring about better results than the alternatives. So the real question, from my perspective, is whether corporate activism is socially beneficial or not.

I'm more sympathetic to Arnold's concerns about *overpoliticization*. Increasing polarization and social tensions are obviously bad for society, and corporate activism robs us of valuable apolitical spaces that could help to mend these civic ruptures. Well, maybe. I'm not sure how much civic healing was ever achieved by shopping, anyway. It's not exactly a community-building activity, and is "cooperative" in only the most abstract sense. So it's probably a pretty minor loss, compared to the independent decline in neighborhood groups and civic organizations.

To take things in a slightly different direction, it might be interesting to think about two different ideals of political neutrality. Arnold seems to have in mind a kind of *procedural neutrality*, in which organizations simply *steer clear* of political disputes. But an alternative ideal would be to aim to promote *substantive neutrality*: protecting ordinary people from undue political encroachment upon their lives.

To see how the two could come apart, suppose that a state government passed legislation requiring that LGBTQ individuals wear a prominent rainbow badge, so that others can more easily discriminate against them. Even if supported by a democratic majority, such intrusion clearly constitutes an objectionable form of political interference in others' lives. Someone *substantively* opposed to overpoliticization could reasonably hope that businesses and other powers in civil society would rebel against it, *precisely to protect individuals from undue political encroachment upon their lives*. Procedural neutrality in the face of such overt oppression would seem more like complicity.

Of course, the practical difficulty is that people disagree about what qualifies as "overt oppression," and what is mere "culture war" that people other than gender studies professors can reasonably ignore. I don't have an answer to this. I just think the issue is more complex than perhaps comes out in Arnold's initial essay.

For what it's worth, I'm inclined to see substantive neutrality as a more appealing standard than procedural neutrality. Reducing conflict is more important than avoiding conflict. If an organization can achieve the former only by taking a stand that itself engenders some degree of conflict (i.e., against unjust political aggressors), then so be it. But I appreciate that this is a risky principle, and easily abused. Unnecessary and intrusive political aggression may be falsely "justified" on the grounds that it is merely "fighting for peace," or to protect the oppressed. In claiming that fighting for peace can be justified, I do not suggest that everyone who *claims* to be fighting for peace is truly justified. The details matter.

Notes

1 For other arguments questioning the orthodoxy that justice always takes priority over beneficence, see Barrett 2022.
2 Reich 2018, 133 mentions a similar proposal that he calls "civil society stakeholding grants."
3 I flag several more questions in a 2019 blog post introducing the idea: https://www.philosophyetc.net/2019/07/charity-vouchers-decentralizing-public.html
4 https://www.givingwhatwecan.org/en-US/funds/effective-altruism-funds

4
DAVIS'S RESPONSE TO ARNOLD

According to what Sam Arnold calls the Default View of philanthropy, people who give have wide discretion in how they give. You can pick what organizations you want to give to, and it's still a morally good thing to do. Arnold develops an argument against this seemingly plausible outlook. First, if you acquire wealth through injustice, then you are not free to donate as you choose. Second, we are all complicit in political injustice. So, we are not free to donate wealth as we choose. The Default View is wrong.

Arnold poses a clear and morally important challenge. As a defender of wide discretion in giving, I do feel the force of his argument. However, I am still inclined to think that the Default View is correct. I will consider each premise in turn.

Wealth through Injustice?

What difference does it make if you gain wealth through injustice? Arnold imagines a case in which you decide to give money to the most effective charity possible, which sounds great until we learn how you acquired the money—by robbing a bank! Arnold holds that *how* you acquired the money constrains *what* you can do with it. Justice demands returning it, rather than giving it to charity.

I'm not so sure. Here's another version of the case. Rather than robbing a bank, suppose you steal a car. Delighting in your good luck as you speed away, you happen upon a small child in the middle of the road who—you guessed it—is in grave danger. The child is in the path of a runaway grand piano, and you only

DOI: 10.4324/9781003508069-6

have time to save the child by putting your new car in between. The car will be totaled, but the child will be saved. Is it OK to save the child? I say: absolutely yes, even if that makes it impossible for you to return the car to the rightful owner, and even if you have no other resources with which to compensate the owner. The reason is simple: It's morally more important to save a child than to remediate a property rights violation.[1]

The moral is that while Arnold is plausibly generally correct that past injustice constrains our use of resources, it's another question about whether injustice constrains giving in the case at hand. Arnold agrees that we are generally morally required to give in the most effective ways, because Effective Altruists are right that the good we can do is more important than our own moral discretion. Granting that thought, isn't saving a life more important than others' projects or property as well?

My own view differs from Arnold's about both kinds of case. I think we have wide discretion in the case of ordinary giving. But I also think it's permissible to crash the car to save a life, or likewise to donate stolen money to save a life. What could explain the difference? As I write in my contribution, I do not believe others have rights to our aid, and so the reasons to save others do not have any *requiring* strength. That is, altruistic reasons don't render other options available to us impermissible. However, our reasons to save another's life have extraordinarily high *justifying* strength. That is, saving a life could justify many things that would otherwise be impermissible.[2] I know Arnold disagrees with me about the first of these claims; he thinks that our reasons to save others can require our action. But I wonder if he also disagrees with the second. Doesn't it seem very plausible that our reasons to save others have exceptionally high justifying strength? If so, then it seems to me that this will make trouble for his first premise. You may be complicit in injustice in getting some wealth, but still justified in donating it—because the justifying reasons supporting donating are so strong.

How Complicit Are We?

Now let's turn to Arnold's case of the Catastrophically Bad Lifeguard. In this case, a neighborhood pool club's manager hires

Billy—the catastrophically bad lifeguard. Not only does he fail to save children from drowning, he actively pushes them into the pool. Despite it all, he isn't fired.

Arnold holds that members of the neighborhood pool are complicit in the injustice. This sounds right, although I think it will matter whether they knew what was going on, whether they put in due diligence to understand the situation, and so on. My central question is about what this case shows about the actual world. Arnold argues that we are all complicit in a democratic state's injustices in the way that members of the neighborhood pool are complicit in Billy's terrible lifeguarding. But for this comparison, it really matters how political states work. Consider a sequel to the case:

> *Catastrophically Bad Lifeguard II*. A terrible lifeguard obsessed with drowning people named Billy conspires with a club manager to be installed as the pool attendant at a neighborhood club. Members immediately object to this obviously disastrous selection, but Billy and the manager threaten to burn down all their houses unless they keep paying their dues. To their horror, neighbors discover that Billy and the manager also control local law enforcement—there is no meaningful ability to individually resist their choices. A few neighbors even decide that supporting Billy and the manager is the best they can do under the circumstances.[3]

Is the political state more like the original *Catastrophically Bad Lifeguard*, or the even more dystopian spin-off? In Arnold's original case, neighbors join the pool club, and other neighbors refrain from joining the club. But none (or anyway, very few) of us have consented to rule by the political state. In the original case, members of the club could reasonably exercise their agency to curb the manager's power. In contrast, individual citizens have effectively zero influence over the political state's exercises of power. This is somewhat less true at the local level, but even there it's mostly true. And in any case, Arnold's illustrative examples are about large-scale exercises of political power: closing national borders, colonialism and conquest, the war on drugs, etc. Far from being responsible for these evils—and to be sure, they are evils—most ordinary citizens are among the victims. Closing borders, for example, is especially terrible for people unjustly kept out, but it also wrongs the rights of association for political society's own members. Neighbors who oppose the swimming pool regime are free from any complicity. But even those who go along with it have a good excuse. Their choices are endogenous to a terrible system.

I think something similar about citizens in a democratic society. Democracy is certainly better than other forms of governance, but the idea that it affords individual citizens the power to curtail the political state's power is an illusion. We are not complicit in the state's injustice. So I also contest Arnold's second premise.

Why Does Democratic Philanthropy Matter?

The two versions of our story do share something in common: Both portray the lifeguard as doing an exceptionally terrible job. I agree with Arnold about his analogous claims against the political state. The history of political rule is a history of injustice. My intuition is that if some agent is doing such a bad job from a moral point of view, the last thing we would want is to entrust them with doing more. Yet Arnold takes the opposite tact. He holds that it's actually morally important for the political state—rather than private persons or institutions—to be the agent tasked with solving social problems. He writes, "I have a duty to care for my son; if you swoop in and do it for me, my son is cared for, but not—one wants to say—by the right person." I accept this kind of thought about one's son, but that's because most parents have very important relationships with their children. In contrast, individuals do not have important relationships with the political state, any more than they have relationships with the Gates Foundation or Mark Zuckerberg. Imagine if someone responded to Catastrophically Bad Lifeguard by saying: "Look, we can't just fire Billy. After all, he's got a relationship with everyone in the neighborhood." To the extent this is true, so much the worse. Billy has done a terrible job! At the risk of mixing metaphors, I'd hire Mark Zuckerberg to lifeguard my pool any day if it meant getting rid of Billy. Likewise, if Arnold and I agree that the state has littered history with injustice, and then we learn that some other agents will try to help people without using any force or coercion, that sounds pretty good to me.

Arnold resists this because he thinks that if billionaires engage in philanthropy without democratic constraints, they will "drown out the voices of ordinary citizens." Here I disagree, and also disagree that it matters. I disagree because I don't think the voices of ordinary citizens are being expressed through democratic institutions. For the most part, I don't even think that citizens' reports of political attitudes are usually expressing beliefs in the ordinary sense.[4] Instead, they are mostly responding to signals from political elites. That means they're more or less repeating back slogans they hear, which doesn't sound much like having a voice of one's own. There is a lot of social scientific reason for pessimism about the prospects for democratic accountability.[5] Of course, Arnold isn't responsible for democracy's problems, and he may have in mind a better functioning conception of democracy. But then it seems like democracy has bigger problems than very wealthy philanthropists.

Second, I don't really understand why it matters. I understand worrying about children drowning, but I don't get the big deal about drowning voices.[6] Think of the injustices Arnold cites. What we want is for people to get an education, or not be put in a cage for crossing a border, or have the resources they need to survive. Who cares who's footing the bill? To paraphrase Brennan's positive view: "What matters is what's effective," not whether it's done by businesses or governments or billionaires. All that matters is getting justice done.

Conclusion

I'll end with an important point of agreement. In his final section, Arnold makes an excellent case that businesses should stay out of socio-political causes. I'm convinced. Overpoliticization is bad for us all. I think Arnold is right in his criticisms of corporate activism.

Notes

1. Moller 2019.
2. For more on the distinction, see Gert 2007; Tucker 2022a; 2022b.
3. Another inspiration for this is Huemer 2012.
4. Claassen et al., 2021; Dias and Lelkes 2022; Ruckelshaus 2022; West and Iyengar 2022.
5. Bisgaard and Slothuus 2018; Kalmoe and Mason 2022.
6. My view is that Brennan's (2016) normative outlook on democracy is a much-needed corrective for democratic theory.

5

ARNOLD'S RESPONSE TO BRENNAN, CHAPPELL, AND DAVIS

Let me start by thanking Brennan, Chappell, and Davis for their incisive and illuminating commentary. I found much to agree with in their remarks—but also much to push back on, as I'll now explain.

My entries explored the "political perils of doing good." Beneficence, I argued, potentially conflicts with political ideals we rightly hold dear. Doing good can undermine justice; doing good can undermine democracy. Or so I suggested.

How should we resolve such conflicts? Suppose you could do something good, but only at cost to justice, or to democracy. Should you do it anyway? You won't find a definitive answer in my entries. My main aim was to show that beneficence potentially clashes with political values. I wanted to *uncover* that conflict, not *resolve* it. I wanted to map the normative terrain, not blaze a path through it.

"Beneficent action potentially casts a dark shadow in the realm of political values," I wrote. That's me *asserting the existence of the conflict* between beneficence and various political ideals. I continued: "That doesn't, of course, mean any given act of beneficence is unjustified, all things considered; but it *does* mean that the question, 'Should I perform this beneficence act?' is more complicated than one might have supposed." That's me *declining to resolve the conflict*.

I once read that there are only two objections to any argument: "oh yeah?" and "so what?" In this spirit, one might contest my argument by *denying that beneficence conflicts with political values*: that's the "oh yeah?" objection. Or, one might concede the conflict, but *deny its significance*: this is the "so what?" objection.

In what follows, I'll defend my argument from both objections. Along the way, I'll respond to many (though not all!) of my co-authors' critiques.

The Conflict Between Justice and Beneficence

Sometimes, beneficence undermines justice.

Suppose you unjustly smash my window, causing $500 worth of damage. "Hey!" I shout. "You owe me $500!" And I'm right: you *do* owe me. You have a moral debt to pay. Now, it just so happens that you have $500 in your pocket. But as you're about to hand it over, clearing your moral books, a wandering troop of Girl Scouts ambles by. "Excuse me," their leader interjects. "We'd like to go on a camping trip that will make us very happy. But we're $500 short. Would you care to donate?"

You face a choice. You can donate, or you can pay me back. Suppose you donate. Was that the correct decision? On one hand, you did something good. On the other, you did it using money that rightfully belongs to me. As I put it in my entry, there's a "moral lien" on your resources. Having wronged me, you're in moral arrears. From the standpoint of justice, you should pay me back prior to engaging in any discretionary do-gooding.

So, there's a conflict here. Beneficence pulls in one direction, justice in another. Giving, even giving well, can be unjust. As Larry Temkin puts this point, "fulfilling one's duties and responsibilities can be at odds with the aim of doing ... good ... with one's resources."[1] That strikes me as important and interesting. But still, a question lingers. How should such conflicts be resolved? Which wins, the right or the good?

As my co-authors rightly stress, the answer depends on the details. How severe is the injustice? And how much good could be done by committing it? Davis asks whether a car thief should sacrifice his new ride to save a child from a runaway grand piano (even though this means not giving the car back to its rightful owner, as justice demands). He says "yes," and I'm inclined to agree; this is a case where the good trumps the right.

However, I don't think the good *always* trumps the right. Better to repay me, in the broken window case, then to make a gaggle of Girl Scouts mildly happier. This suggests that there are at least *some* cases in which, to quote Temkin:

> personal responsibility for one's actions ... prohibit[s] people from using their resources to [promote the good], because others now have a rightful claim on those resources; so, in the morally relevant sense, the resources are no longer "theirs" to do with as they see fit.
>
> *(288)*

In Chiara Cordelli's pithy formulation, these are cases in which "debtors cannot be choosers."

So, (1) justice and beneficence can conflict; (2) sometimes justice deserves to win; but (3) sometimes it doesn't. My critics are right that I should have more clearly acknowledged (3).

Are We Complicit in State Injustice?

If you're on the hook for an injustice, you are under a defeasible obligation to use your resources to rectify that injustice. (To say that the obligation is "defeasible" is to say that it may be outweighed by stronger considerations.) That's one way to put the upshot of our discussion thus far. That takes us about halfway through my first entry.

In the second half, I extended the argument by claiming that basically everyone reading these entries *is* on the hook for serious injustices. This is because, being citizens of democratic states, they're *complicit* in injustices committed by those states (of which there are many). In short, we're on the hook for the many governmental misdeeds committed on our behalf and in our name. Hence, there are *lots* of moral liens on our resources—lots of unpaid moral debts constraining our donative discretion. Or so I argued.

This argument didn't sit well with my co-authors. None of them bought it. In particular, they rejected the idea that citizens of democratic states are complicit in their government's injustices. I suggested, somewhat feebly, that citizens are a bit like members of a pool club that hires a maniac for lifeguard, and continues to employ him despite his crimes. In a similar way, we citizens elect officials who then do horrible things in our names—all without facing any penalty. We voted these bums into office; we pay their salaries; aren't we complicit in their wrongdoing?

As my co-authors persuasively show, things aren't so simple:

- As Chappell points out, supporting the government may be the best of a bad set of options, in which case it doesn't seem wrong. Imagine a version of the pool case where Billy, bad as he is, looks like a choir boy next to all feasible alternatives. In such a scenario, supporting Billy is actually the right thing to do.
- Moreover, many people *don't* support the government. As Brennan writes, "in democratic countries, many people vote against injustice, join reform parties … engage in activism, and so on." Since they "do the best they can, given the background constraints," surely these citizens "are not complicit"?
- Finally, Davis notes that "individual citizens have effectively zero influence over the political state's exercise of political power." But if they're powerless, they can't be complicit. How could they be on the hook for the state's crimes, when nothing they do could curtail those crimes?

These are all excellent points, and together they show that I got a bit over my skis in making this "we're all complicit" argument. I happily retract my claim that all ordinary citizens are complicit in their state's injustices.

Instead, I put forward a modified claim. Actually, two. First, as argued above, *if one is responsible for an injustice, then the victims of that injustice have a defeasible claim on one's resources, sufficient to rectify the injustice.* If you smash my window to the tune of $500 damage, then you should probably pay me $500, even if this means forgoing welfare-improving expenditures. (Within reason: this obligation, again, is merely defeasible. If you can cure cancer for $500, please do so rather than repaying me.) Second, *to the extent that citizens are complicit in their state's injustices, of which there are many, then they bear some responsibility for those injustices—in which case, see the first claim.*

Notice that this second claim is a conditional: it says *if* citizens are complicit, *then* they are responsible. It doesn't assert that citizens *are* complicit; it merely says what follows if they are. Accordingly, it sidesteps my co-authors' quite persuasive objections to my original, overly-ambitious argument.

At core, my message about justice and beneficence is this: justice *can* constrain the pursuit of the good; in particular, if you owe someone restitution for wrongful conduct, then there's a moral lien on your resources; depending on the circumstances, you may be obligated, all things considered, to repay your moral debt *prior* to using those resources otherwise.

Brennan asks what any of this has to do with EA. Well, for starters, Sam Bankman-Fried may be morally required to make his defrauded investors whole before, I don't know, funding research into X-risk.

Elite Philanthropy and Democracy

Call me idealistic, but I don't relish living in a country where "outcomes of common concern ... are overwhelmingly responsive to the preferences of the rich." That's why I'm skittish about elite philanthropy. When ultra-wealthy donors reshape the world to their liking, I can't help but think: this seems disturbingly plutocratic. What about democracy? What about political equality? Aren't these important political values, and aren't these under siege in a world where billionaire philanthropists can remake the social world to their liking?

As explained above, my main intention was not so much to claim that democracy always trumps beneficence, but instead to establish that the two potentially conflict. Philanthropy, especially of the elite type I discuss, threatens democratic values, especially political equality. That's the core claim of my second entry, and as far as I can tell, none of my co-authors challenge it. They all grant, at least implicitly, that big-money, policy-shaping philanthropy gives donors vastly more influence than ordinary citizens. Indeed, how could Davis, Brennan, and Chappell *not* grant this point? It's just obvious, surely, that Mark Zuckerberg had more say over Newark school policy than some random Newarkian.

Here we reach the "so what?" objection I mentioned at the start of this chapter. My interlocutors admit that elite philanthropy undermines democracy—but ask: *who cares?* Davis confesses that he "[doesn't] really understand why it matters" whether outcomes emerge from an egalitarian process or a plutocratic one. "All that matters is getting justice done," he writes; who cares *who* does it.

Brennan, author of a famous (and excellent) book called *Against Democracy*, shares Davis's pragmatic outlook. In his book, Brennan contends that democracy is just a tool. Like a hammer, it should be evaluated instrumentally. Does it get the job done? Does it produce better results than alternatives? If so, great; let's use it. If not, let's not get too precious about political equality and the like. Democracy isn't a poem; it's not something to be admired for its beauty, or how it makes us feel, or its intrinsic qualities. All that matters is how well it works.

Chappell agrees. "I'm not sure how much we should care about procedural unfairness," he writes. Would it be *unfair* for an omniscient, benevolent dictator—God, say—to take the reins of power? Like Davis, Chappell finds this question inscrutable: "I'm not even sure what the question is asking." All that "really matters" is whether this dictator's rule "would be an improvement."

"What's missing from Arnold's argument," writes Brennan, "is the essential piece." (That's a bad piece to leave out!) "We need evidence that democratic control [over philanthropy] would, in real life, be superior to billionaire philanthropy. Without such evidence, we have no reason to endorse his conclusions or even care about whether billionaire giving is undemocratic."

Respectfully, I think my critics are missing my key point. I'm not claiming that democratic philanthropy would yield better results than elite philanthropy. I'm not complaining that Mark Zuckerberg is a bad school superintendent. For all I know, he's an educational-policy wizard. *Still, he's just one guy.* Why should *he* determine Newark school policy? Isn't it unfair for him to get so much influence over an important issue of common concern? Nobody elected him School Policy Chief.

In any self-respecting democracy, important issues of common concern belong on "the democratic agenda." Such issues are supposed to be resolved through a process that's not utterly dominated by the rich. Otherwise, we live in a plutocracy, not a democracy.

Billionaire philanthropy, no matter how effective, comes with a significant normative downside: it undermines political equality.

My critics claim not to understand the points I just made. They claim not to get what political equality is, or why it matters. Honestly, I find that hard to fathom. Notice that I'm not claiming that political equality is the *only* thing that matters. I'm not saying: democracy or bust. I'd rather solve big problems undemocratically (via billionaire largesse, say) than not solve them at all. But if I had my druthers, we'd tackle these problems in a non-plutocratic way. We'd solve them through a more-or-less-egalitarian process that's not utterly dominated by a tiny economic elite.

Put it this way. I like solving important social problems. I'm in favor of that. So are my critics. So that consideration gets factored out of our dispute. Where we differ is on the question of democracy versus plutocracy. I submit that, *other things equal, democracy is morally preferable to plutocracy*. So, if we could solve our problems pretty well via Zuckerbergian, conditional philanthropy, or pretty well via a democratic process that gives all affected parties a roughly equal say, I'd prefer the latter.

Corporate Activism

My interlocutors agree with my conclusion concerning corporate activism—namely, that corporations should largely forbear from weighing in on controversial socio-political topics.[2] Accordingly, I won't do anything here to defend that conclusion. Instead, I want to expand an argument I make in my third entry, an argument that may put pressure on my critics to concede—if only grudgingly—the value and importance of political equality.

Activist corporations have two main tools at their disposal. They've got a megaphone through which to blare their message. And they've got a veto button with which to override disliked social policy.

Let's talk about this veto button. Suppose a state's duly elected legislature passes some law that Globocorp doesn't like. In response, Globocorp threatens "capital strike" *via* boilerplate press release:

"As a values-driven, forward-thinking company, Globocorp is extremely disturbed by the legislature's actions. This new bill goes against everything Globocorp stands for. Accordingly, we are reconsidering our planned $100 million expansion in the state. We may cancel it outright, or we may move it to a different state. The next few weeks will be crucial. Will the legislature do the right thing, and create a climate consistent with Globocorp's values? We'll be watching closely."

Legislators, in response, freak out; the state's prosperity depends on Globocorp! If Globocorp leaves, the economy will take a nose dive, and voters will punish the politicians responsible. So, out of pure political self-interest, the legislature really has no choice but to give in. The bill's main backers, tails tucked firmly between their legs, hold a press conference with Globocorp officials. "We've found a compromise that satisfies the business community along with the great people of this state," they say. "It's a win-win. We look forward to partnering with Globocorp today, tomorrow, and for years to come."

Of course, this press conference is total B.S.; the "compromise" isn't a win-win; it's a rout. Armed with the threat of capital flight, Globocorp has strong-armed the democratically elected legislature. It enjoys a *de facto* veto over the state's policy; by threatening to head for greener pastures, it can smother disliked changes in the cradle. Democratic majorities, therefore, can enact economic policy *only with Globocorp's blessing*.

Now, this is just an idealized sketch, but something very similar happens all the time in the real world (if with a bit more complexity and nuance). As Rebecca Henderson, a professor of business at Harvard Business School notes, "today's firms have enormous power to influence governments if they choose to use it."[3] And use it, they should—or so she argues. She wants corporations to make a habit of threatening disinvestment in response to disliked policy change. They should dangle the Damoclean sword of "capital flight" over the heads of backward states like Indiana, or North Carolina, or Georgia, promising to repay unprogressive policy in the coin of economic ruin.

"Nice economy you've got there," Globocorp should, on Henderson's view, say to the elected representatives of this or that benighted, backwater province. "Hate to see something happen to it. How about you *don't* do that socially conservative thing that, for some unfathomable reason, the electorate wants you to do?"

Here's my point. Maybe you like Henderson's politics, or maybe you don't. Maybe you side with Globocorp on the merits, or maybe you side with the state. Put that aside. Isn't there something deeply disturbing about Globocorp's power? Isn't something important lost when corporations can override democratically chosen laws?

Indeed, there is. Globocorp's veto power violates political equality. It undermines democracy. We're supposed to enjoy roughly equal power over important issues of common concern; or at the very least, we're not supposed to live in a society where everything important gets decided by the most economically powerful actors.

I submit that political equality matters, and that it matters independently of results. As evidence, I present the case of the corporate veto. If you, like me, are disturbed by this case, then it must be because you share my intuition that political equality counts for something, that democracy is, in some way, better than plutocracy.

I don't think my co-authors would say the same. Given their wholesale rejection of political equality as an independent value, the only question they can ask of Globocorp's action is: "Well, was Globocorp right on the merits, or not? Did its veto produce better results than the law would have?" I agree that matters, but I deny that this is all that matters.

We want the right results, but we also value reaching them in a particular way: namely, through a democratic process that respects our status as civic equals.

Notes

1 Temkin 2022, 287.
2 I don't think they share my *reasoning* for this conclusion, however. My argument, roughly, was that corporations should not engage in activism because that would give them political-equality-violating levels of influence over issues on the democratic agenda. Since my co-authors aren't fans of political equality, this argument necessarily leaves them cold.
3 Henderson 2020, 44.

PART II
Effective Altruism and Regular People

Jason F. Brennan

6
EFFECTIVE ALTRUISM AND REGULAR PEOPLE

Jason F. Brennan

Where I'm Coming From

Consider these basic insights from microeconomics and moral psychology:

1. Think on the margin. Consider what good the next unit of time, effort, or money will do.
2. Think strategically. Consider how others will react to what you do, or how they will act when they try to predict what you will do.
3. Think about costs. There is no such thing as a free lunch, even (especially?) when it comes to ethics and justice.
4. Think about opportunity costs. Ask what the best alternative is.
5. Think about diminishing returns and increasing costs. For most inputs, each additional unit does less good than the previous. When the marginal costs exceed the marginal benefits, it's time either to innovate or stop.
6. Look for the unseen, not just the seen. When making decisions, don't just focus on the most obvious effect for the most directly affected in the short term. Consider the less obvious effects for everyone over the long term.
7. Don't confuse intentions with results. Meaning well doesn't automatically mean doing well; meaning badly doesn't automatically mean doing badly. A good-intentioned act can be a disaster and a selfish act could be a blessing.

DOI: 10.4324/9781003508069-9

8. Actions have unforeseen consequences. We cannot go on our gut feelings. We have to check to see what our decisions really do.
9. People are people. There are slight differences in motivation among who decides to become a surgeon, a CEO, a nurse, a teacher, a car salesperson, or a politician. Still, people are predominantly selfish and only secondarily altruistic. A few are malevolent much of the time, and many become malevolent in particular, predictable situations. So, when designing or working with institutions—including charities and NGOs—we should consider what realistic people are likely to do rather than what we hope idealized people would do.
10. People have predictable quirks and biases, such as blind spots and a tendency to conformity, which prevent them from being as good as the plan to be. We can work around these somewhat.
11. Incentives matter. We get more of what we reward and less of what we punish.
12. People tend to make smarter choices when they bear the costs and benefits of their actions, and stupider choices when they can externalize the costs and benefits of their decisions onto others.

As a practical, substantive program, rather than an abstraction, I see Effective Altruism as a set of related theories and practices that apply ideas like these to beneficent action.

To me, Effective Altruism seemed like a natural outgrowth of the kinds of ideas and arguments I already use in political philosophy. For instance, take the subfield of democratic theory. As a grad student, I avoided it, thinking it read like high-falutin' nonsense that had nothing to do with actual democracy. Later I realized these pervasive flaws were a reason to work in the field, because that meant there was work to be done.

Most democratic theorists rely on idealized models of how democracy ought to work. Their normative theories of democracy don't apply to actual democracy. In contrast, I approach democratic theory by first examining how political institutions in the real world function, and then asking which institutions are best in light of the realistic constraints. This work is harder—I joke that ideal theory is like arithmetic and non-ideal theory is like differential equations—but it tells us what to do rather than how to daydream.

As a philosopher working a business school, I teach classes on human behavior, institutions, and organizations. Many of my classes, such as "Managing Flawed People," are about what makes people tick, assessing why they make bad choices or make decisions that fall short of their noble intentions, and then planning how to arrange one's personal and social environment to make it more likely one does the right thing. Effective Altruism is a set of tools in this vein.

The three questions I chose below reflect these ideas. First, I ask whether Effective Altruism is merely an applied version of utilitarian moral theory, or whether instead it's compatible with a wide range of moral views. I argue utilitarians are not ideologically committed to Effective Altruism and Effective Altruism is not grounded in utilitarianism. Second, I inquire whether we can and do exercise beneficence through things like normal work and investing, rather than giving and volunteering. Third, I ask why so few people are Effective Altruists, and what, if anything, can be done about it. I ask this third question in part because I don't know the answer. I am skeptical that we can induce the typical person to behave much better.

Question 4: Can We Endorse Effective Altruism Without Endorsing Utilitarianism?

Effective Altruism might seem like little more than rebranded utilitarianism, or like utilitarianism applied to giving. Utilitarianism says we ought to do the most good. Effective Altruism advises us how to do the most good. Same thing, right?

Of course, utilitarians should endorse effective giving, but Effective Altruism is more than some abstract commitment to effectiveness. Rather, it has a substantive methodology for assessing charitable endeavors plus a substantive view of which endeavors are worth pursuing. Sure, Effective Altruists can disagree among themselves, but not everyone committed to effective altruism is an Effective Altruist per se. It's coherent for a utilitarian to say, sure, I want to be effective in giving, but I reject most of the substantive things Effective Altruists say, and thus don't consider myself an Effective Altruist. So, utilitarianism is by definition committed to effective altruism but not by definition committed to Effective Altruism.

Still, Effective Altruists often define or summarize their theory in ways that sound utilitarian. Roman Duda at *80,000 Hours* says that effective altruism is about "using evidence and reason to figure out how to benefit others as much as possible, and taking action on that basis."[1] William MacAskill, in *Doing Good Better*, says that his two primary questions are, first, how to ensure that when we try to help others, that we "do so as effectively as possible," and, second, how to ensure we avoid harm and instead have the "greatest positive impact we can."[2]

But even these statements needn't commit Effective Altruists to utilitarianism. Effective Altruism doesn't, and isn't meant to, answer these questions:

1. How much self-sacrifice do I owe others? What is the minimal amount of good I must do for others? How many hours, how much effort, how much money must I give? How demanding is the morality of us?
2. Should I be impartial to everyone's welfare? How much personal prerogative do I have?
3. Is it permissible to harm one innocent person to help another, if the gains to the winner exceed the losses to the loser?

Utilitarians often answer these questions in ways which Effective Altruists need not endorse. Some utilitarians argue that we should (at bottom) be impartial among everyone's welfare.[3] Some utilitarians say we should spend our lives doing the most we can for others; it's impermissible to live high while people die.[4] Utilitarians often say it's permissible to sacrifice the few for the sake of the many so long as the utilitarian calculus comes out right. Effective Altruism per se is not committed to such claims.

Instead, Effective Altruism is a set of substantive intellectual tools meant to help us transform our benevolence into beneficence. Its goal is to ensure that when we try to help others, we succeed. Consider some substantive Effective Altruist ideas and questions:

1. A charity consumes inputs (effort, money, time, resources) and produces outputs (lunches served, deworming pills distributed, etc.), which in turn produce various effects (life-years extended, greater height, IQ boosts, etc.). How can we tell whether the value of the outputs exceeds the value of the inputs?
2. We should think on the margin. The Red Cross might do more good than some smaller charity, but the Red Cross might have sharply diminishing returns while the small charity might have room for funding. Giving the small charity $1000 might do lots of good, while giving the Red Cross another $1000 might do nothing.
3. For the same reasons, we should consider placing our money and time in overlooked and neglected issues.
4. Giving might do more good than volunteering. Giving money to a charity might do more good than working for it.
5. The "poor" in the West are quite rich by world standards. Buying a laptop for a teenager whose parents make $22,000 a year in the rural US might probably does less good than deworming a child whose parents make $1000 a year in rural India.
6. Many charities mean well, but their interventions fail or backfire. Many consume so many resources that, on net, they do no good.
7. Things other than "charity" do good for others.
8. Social enterprises and even regular businesses are sometimes equally or more effective at delivering the goods and services we tend to think governments and charities are supposed to provide.
9. We have psychological biases toward doing what sounds good or looks impressive to others, rather than what actually helps. We can identify and overcome these biases.

And so on. You don't have to be a utilitarian to endorse these ideas or think these are questions worth asking.

Consider an analogy: Many magazines evaluate various cars in terms of handling, speed, reliability, and cost. For instance, an author in *Road and Track* claims the Honda Accord Touring compares decently to BMW and Audi in terms of handling and interior quality, but at far lower cost.[5] *Car and Driver* has listed the Honda Accord among its "10 Best" for 37 consecutive years.[6]

Imagine a critic reading these reviews and saying, "Well, I'm not a utilitarian. I don't think I have any obligation to buy the *best car for the dollar*. I can feel free to buy a lousy car, like a Nissan Versa hatchback, or a car that has low bang for the buck, like a Jaguar XF." This objection falls flat. You need not be a utilitarian to think that *Consumer Reports* gives useful advice about which cars to buy or avoid. You don't have to hold that you are duty-bound to buy the best car for the dollar to think the advice these magazines offer should help shape your choices. Similarly, you don't need to be utilitarian to think Effective Altruism gives useful advice for choosing careers or charities. That a car is unreliable, has bad gas mileage, and drives like a fog cloud are reasons to avoid it. That a charity like Scared Straight causes $200 in social harm for every dollar it spends is a reason to avoid it.[7]

Most people have moral beliefs, but they do not subscribe to something as robust as a moral theory, per se. Commonsense ethics holds people have what philosophers call an imperfect duty to help others. This imperfect obligation of beneficence becomes stronger the more easily we can help. Still, in commonsense moral thought, with some special exceptions, it's largely up to us—as a matter of practice wisdom that cannot easily be codified—when and how to exercise our obligations of beneficence. Many high-level moral theories, such as Kantianism or Aristotelian virtue ethics, concur with this commonsense but wishy-washy view.

Even people with such wishy-washy views can endorse Effective Altruism. Indeed, to least some degree, they have to. Suppose you hold the commonsense view that you should sometimes help people. You agree that the more time and money you have to help, the more you should do. You aren't really sure what the cut-off is—that is, at what point you've done enough good that any additional volunteering or charitable giving counts as above and beyond the call of duty. Nevertheless, the imperfect obligation of beneficence is an obligation to actually help. If you donate 20% of your income to a harmful charity like Scared Straight, or a charity that does no net good, you haven't thereby discharged your imperfect obligation of beneficence. You tried to help, but you failed.

Imagine I wanted to feed the hungry, so I placed loaves of bread inside bottles and dropped them in the ocean, hoping shipwreck survivors would find them. Alas, no one does. Regardless of whether I dedicated 1% or 100% of my income to this endeavor, and regardless of my intentions, I haven't yet acted beneficently.

So, at the bare minimum, even someone with wishy-washy commonsense views must endorse what we might call *Minimally Effective Altruism*. An obligation to help is an obligation to help, which means that any giving or volunteering they do must at least do more good than harm, and produce more value than they consume.

Further, cases like this help us understand something implied by commonsense morality, but which people often miss. The obligation to do good is an obligation to do good, not an obligation to *sacrifice*. Asking whether we should donate 1%, 10%, or 25% of our income, or volunteer 1, 10, or 25 hours (a month? a year?) is *beside the point*. Saying we should give in proportion to how much we have to give and how easy it is to give is *beside the point*. These are *costs*, not outputs. We should shift moral discourse away from how much we should *lose* in giving instead toward how much we are obligated to help. Giving is instrumental to helping, not an end in itself.

Some philosophers argue that commonsense morality implicitly commits us to choosing the *most* effective course of action. For instance, to modify some examples from Theron Pummer and Christopher Freiman,[8] imagine you decide to feed the hungry with your leftover sandwich. You can give it to either to a rich student, thus saving her from walking to the cafeteria, or a starving child, thus saving him from dying. It seems wrong, even in commonsense thought, to give it to the student rather than the child. You don't have to be a utilitarian to agree.

That seems right, but it's unclear that such cases commit us to concluding that once we decide to help with certain resources, we are obligated to do the *most good* we can with those resources. One reason why not is that considerations other than pure beneficence bear on our giving. When I donate money to my alma mater, I act partly on a principle of reciprocity rather than pure beneficence.[9] When I volunteer to play guitar in high school musicals, I act on a principle of community rather than pure beneficence.[10] What Plummer and Freiman might show is that (1) when the *only* relevant value consideration is need, and (2) when it's obvious which choice meets the most needs, we should (3) meet the most need. But in the real world, other value considerations are often at play, and it's often not obvious what does the most good.

Indeed, Effective Altruism itself says we are supposed to be strategic, think on the margins, and do cost-benefit analysis. This means that even when in assessing charities, we have to consider whether it's worth our time, money, and effort to do additional investigative work.

Question 5: Is Beneficence Confined to Charity and Volunteering, or Can We Exercise It in Other Domains, Including Business?

In the received wisdom, people tend to think beneficence is about charity, volunteering, or spontaneous acts of heroism. Beneficence is about giving away money and time, or risking your life to save the child in the burning building. Beyond that, at first pass, they tend to think the available ways to exercise beneficence are limited.

Upon further reflection, some might concede that (some) government agencies provide goods and services that meet people's needs. Most people don't then conclude that paying taxes toward welfare programs or public goods counts as beneficent—after all, that was Ebenezer Scrooge's argument for not helping the two men raising money for charity. (Scrooge, like many modern egalitarians, says that he discharges his duties by paying his taxes.) Still, many people regard certain government agents as holding beneficent jobs—as exercising beneficence through their work.

Upon reflection, many people also regard workers in certain *care* professions as beneficent or acting beneficently through their work. They praise nurses, medical doctors, and teachers, even though some such people are paid extraordinary well. They are selective in their praise: A plastic surgeon won't normally be praised as beneficent (except in special cases, like reconstructing a toddler's face after a pit bull attack), but a pediatrician would be.

Effective Altruists advise us to look beyond our gut prejudices and to consider our impact instead. Many Effective Altruists then argue that if we want to help others, it is often more effective to *earn to give* than to work directly in a care, social work, charity, or social justice profession. A Bain Capital executive who donates a third of her income to Against Malaria might have a greater marginal impact than someone who works directly for Against Malaria.[11]

Further, some entrepreneurs, business ethicists, management theorists, and Effective Altruists argue for expanding the list of things we consider beneficent even more. In particular, they accept that social enterprises or social businesses are in some cases viable or superior alternatives to charity. Social businesses aim to use for-profit market mechanisms to supply the goods or services traditionally supplied by charity or government. In some cases, the argument goes, charities and governments suffer from information or organizational problems which a social business can overcome. If one finds this argument persuasive, one might concede that running or working for a social business can qualify as beneficence.

What I call the *central philosophical problem of charity* is that, in the absence of a profit-loss accounting mechanism, it is difficult to judge whether a given charity is creating or destroying value. There is often no clear or easy mechanism to commensurate the value of their outputs with the cost of their inputs. Listing inputs and outputs is not enough. We have to put them on a common scale.

Contrast that with business. In a market, the price of goods and services is determined by the forces of supply and demand. In turn, these forces emerge from all market participants' individual knowledge and desires. Prices thus contain information about how others value things and what tradeoffs they make. In a competitive, free, and unsubsidized market free of significant externalities, for a company to profit, it must transform inputs that people value at one level into outputs people value more. Profit is possible only when businesses create value for other people. So, in business, judging whether you created value is easier.

There are complexities here—thanks to, say, differences in purchasing power among the poor and rich, the degree of profit does not always indicate degree of goodness. Nevertheless, it remains *easier* to judge whether a business creates value than whether a charity does. A profitable business in a competitive free market, free of externalities, is a value-creator. Alas, there is no analogous mechanism to judge charities.

What I call the *central management problem of charity* is that a charity's "customers"—that is, the people who *pay*—are not their intended beneficiaries—that is, the people who *receive* the charity's goods and services. The charity's customer is the donor. A charity keeps the lights on by pleasing donors. However, donors are often wrong about what works, or what's good for the beneficiaries, or have pet projects. NGO managers face perverse incentives to please donors even when donors are wrong or undermine the NGO's mission.

In a free market, a business stays open by making customers happy. Chipotle survives only when enough people are willing to pay for their burritos. This forces Chipotle to serve their customers. It profits only if its customers profit. But charities can remain open even if they consistently harm their intended beneficiaries, so long as donors keep writing checks.

Social entrepreneurship intends to overcome these problems. Social businesses aim to help the worst off or those suffering from injustice and deprivation, but use market mechanisms to help identify, test, and ensure that they add more value than they consume. While a charity tries to help needy people by making them *beneficiaries*, a social business tries to help needy people by making them *customers*. A charity must convince donors that its charitable actions are worth the price the donors pay. A social business must instead ensure that its needy customers *profit* by buying the social business's goods and services. Social businesses survive only when their outputs are higher value than their inputs. This gives social business an epistemic advantage over charity—the fact they make a profit is evidence they create value for others.

Social businesses do not suffer from the central philosophical or management problems of charities.

Many people agree, for these reasons, that social businesses can be effective alternatives to charity, and so will count working for social businesses as possible forms of beneficence. But I think we should expand the list of things we count as beneficent even more than that.

Consider that the duty of beneficence is a general obligation to help others.[12] To act beneficently is to aid others, improve their welfare, and meet their needs. The duty of beneficence is an obligation to provide aid or work to the benefit of strangers and people at large.[13] The more we reflect, the more we realize that beneficence can be exercised outside the stereotypical actions of donating or volunteering. Just as courage is not limited to the battlefield, so beneficence is not limited to the donation jar. Beneficence can be exercised anywhere.

I think the typical person, working the typical job at the typical employer, already acts beneficently *by doing that job*. They already discharge at least some of their obligations of beneficence through such work. While earning to give can be noble, most of us already *give by earning*. Most of us do far more for others through our productive work than through charity, not merely because we are bad at selecting effective charities or because we give too little to charity, but because regular work does more good for others than we realize.

As I discussed in the previous question, the point of beneficence is to help others. Sacrificing time, money, and resources is not a goal, but at best a means. The point of beneficence is not for the giver to lose but for receivers to win.

Imagine an evil demon placed a curse on charities, such that only a minority of them do much good or create much more value than they consume. But imagine that an angel countered by placing a blessing on for-profit business, such that most of them do lots of good for others. In such a world, if you wanted to act beneficently, you'd be cautious about donating to charity but could accept most jobs at most businesses without much worry.

What if our world turns out to be somewhat like that, minus the demons and angels? Effective Altruists have already shown that many charities are ineffective. Still, many readers (including Effective Altruists) might resist accepting that normal, productive work counts as an exercise of beneficence. Should they?

Imagine a friendly genie offers you three choices:

A. He will donate $100 to charity. The charity does $120 of good for others.
B. He volunteers for one hour at a charity. The cost to him of his labor is $100. His work does $120 of good.
C. He works for one hour at a factory. The cost to him of his labor is $100. His work does $150 of good for others.

Here, imagine that he's commensurated all the prices, taking into account the differences in value his outputs might have for different people, and removed any biases which result from different levels of effective demand from the rich and poor. If these are your only choices, you should choose C. Or, at the very least, C is a good choice for those who want to help. Otherwise, you are fetishizing how you help over how much you help.

One reason people resist this is that they think for-profit work is selfishly motivated. But even if it partly or mostly is, that doesn't imply that it's not beneficent. Consider an analogy to charity. Suppose Bob gives $200 to a highly effective charity, not because he cares to help people, but to impress his girlfriend Jane. In contrast, Charlie means well but is misinformed, so he donates $200 to a counterproductive charity. Charlie is more *benevolent* than Bob, but Bob is more *beneficent* than Charlie. Bob actually helps; Charlie means to help but doesn't.

More generally, we need to be careful to distinguish the quality of a person's motives from the quality of their actions. As I will discuss below, even when it comes to charitable giving, people's selfish desires are stronger than their desire to help. Still, that doesn't in itself mean their actions are not beneficent. It instead shows they are not as benevolent as they think they are.

With that, consider a variation on C:

D. The genie works for one hour at a factory. His work does $150 of good for others.

If you only care about helping others as much as possible, you'd be indifferent between C and D but prefer either to A and B. However, once you admit your welfare matters too (why shouldn't you?), then D is better than C. If our goal is to help people as much as possible, then the giver's welfare matters too. A benevolent third party would prefer D to C, because D is Pareto-superior.

By analogy, consider a highly effective charity, such as Sightsavers. Imagine that an angel cast a blessing, such that every time you donate $500 to them, they not only do the good for others they normally do, but you also magically get $600 back. This magic spell wouldn't give you less reason to donate to them, but more. Remember: The goal is to help people, not to sacrifice yourself for others. If helping people also helps you, great!

Effective Altruism advises us not to judge charities, governments, NGOs, or social businesses by the *intentions* of their workers or founders. But neither can we judge for-profit businesses by their intentions. Further, the same mechanisms that make social businesses serve their customers appear in for-profit business. The same epistemic advantages do as well.

The typical individual transaction on the market benefits both parties. But that understates the good business does, because business as a systematic whole has major positive externalities. Economics tells us that the systematic effect of business trade and investing has been to create background conditions of wealth, opportunity, and cultural progress. We each benefit from the positive externalities created by the market's extended system of social cooperation. We are engaged in networks of mutual benefit, and we benefit from other people being engaged in these networks. When we work in business, we help create, sustain, and improve these networks of mutual benefit. As a result, the overwhelming majority of people are now vastly wealthier than they otherwise would have been. This makes it easier for us to realize our disparate conceptions of the good life. It expands the options available to us. We live longer and better.[14]

David Schmidtz says, "… any decent car mechanic does more for society by fixing cars than by paying taxes."[15] If a person makes, say, $2 million over their lifetime through productive work, it's not as though that person simply took $2 million from others. The person gets $2 million because they did more than $2 million's worth of good for others. It is not only unfair, but innumerate, to say to that person, sure, you worked a lifetime providing valuable services for others, services which make all of us better off, but what have you done for us?

I am not arguing that markets and trade are perfect. Nor do I make any claim here about whether government should sometimes intervene to correct market failures. Further, I will happily admit that, on the margin, many instances of charity will do far more good than many instances of trade or investing. But all this applies to government, medicine, teaching, volunteering, and charity as well.

We don't usually reserve the word "beneficent" only for actions which *maximize* the good done for others. A donation to the local food pantry does far less good than an equivalent cash donation to Evidence Action or Sightsavers, but both actions are beneficent. By extension, buying an Indonesian-made guitar might do less good overall than curing someone of blindness through Sightsavers, but both actions are beneficent.

An Effective Altruist would say that if our goal is to reduce malaria deaths, we shouldn't fetishize anti-malaria pills over mosquito nets. We should pick the intervention that actually works. But this generalizes further. What matters is what's effective, not whether helping comes in the form of volunteering, charity, government, social business, or regular old business.

Effective Altruists note that, on the margin, donations to highly effective charities can do tremendous good in the short term. Still, we have no record of charity solving the problem of extreme need.

How is it that some people are rich—and thus in a position to give—in the first place? In 1900, 95% of the world was destitute, living in what we'd now consider extreme poverty. Today, fewer than 10% of people live in extreme poverty. In 1950, Japan, South Korea, Hong Kong, Singapore, and Taiwan were very poor. Today, in 2024, Japan, South Korea, Hong Kong, Singapore, and Taiwan are very rich—indeed, the average person in Singapore is now richer than the average American. Why?

It's not as though Japan and Korea became rich because people followed, say, Peter Singer's advice, and donated 75% of their incomes. Instead, they became rich precisely because people ignored Singer's advice. Over the past 60 years, people in already rich countries bought toys, transistor radios, stereos, video game consoles, VCRs, DVD players, Blu-Ray players, smart phones, automobiles, electronics, and a wide range of other morally insignificant luxury goods they didn't need from these countries. As a result, hundreds of millions of people were liberated from poverty and joined the ranks of the rich.

A similar point holds for investing money, rather than spending it. Investing doesn't stop people from starving today. It makes it so that people don't need charity in the future.

There is a genuine moral tradeoff here. The world doesn't play fair; it makes us choose between feeding the hungry today or ensuring people can feed themselves tomorrow. Someone like Peter Singer might suspect I am trying to go *easy* on average people, by telling them, no worries, your investments and trade do just as much good as your giving. On the contrary, I am being *hard* on all of us, because if we want to help others, it's far from obvious what the optimal mix of trade, investment, and giving is.

Question 6: Why Are People So Bad at Charity and What Should We Do About It?

Consider why people buy BMWs. Yes, they have nicer interiors and drive better than more pedestrian cars, like a Toyota Camry. But part of the goal is to signaling to others that one has good taste, is a sporty and fun person, or has money. BMW owners can intuit this for themselves: Imagine Toyota made a car that was identical in all respects to your BMW, except for a big Toyota badge. How much more would you be willing to pay for the BMW badge?

Signaling is about communicating information through indirect means. We often engage in signaling when direct communication is unreliable or not trustworthy. For instance, if a job applicant says, "I am smart and perseverant," the recruiter has no reason to believe it. But if the job applicant has a B.S. in physics from MIT, this signals to the recruiter that the applicant is smart and perseverant.

Signaling is ubiquitous. Consider a prime example: engagement rings. We offer expensive engagement rings not just because of tradition or because we were manipulated by DeBeers' marketing. Rather, rings help solve an information problem. Talk is cheap. Anyone can say, in the deep dark night by the dashboard light, that they will love you to the end of time. But potential marriage partners need to know how committed the other person is—and also how committed they themselves are. Giving a diamond ring says, in effect, "I love you so much that I am willing to sacrifice two month's salary on an expensive trinket! Also, the fact that I can afford this trinket is evidence of my financial stability."

Now consider charitable giving or volunteering. When we give or volunteer, we might have all sorts of motives. We might genuinely want to help. We also might simply *enjoy* the work. (I have volunteered to perform guitar for various groups, but I do it partly because it's fun. Notice that I'm not, say, painting theater sets, which I hate doing.) We also might do nice things because we want others to think we are nice. We want others to believe our hearts are in the right place. And so on. It's an empirical question what actually motivates people.

Fortunately, people have done that empirical work. Unfortunately, the results are depressing. We have good reason to think that most people are more concerned with looking good than doing good.

For instance, consider "voluntourism." Some unskilled high school student spends $10,000 to travel to a remote part of Peru, where she spends a week trying to build a school. She takes lots of pictures, which she posts to great approval on social media. She writes about the experience in her college essay, which helps her get admitted to Brown. But if she really wanted to help, rather than look helpful, it would have been far better to spend that $10,000 hiring skilled local carpenters to build the school. It's not she's entirely selfish, but for her, being helpful is secondary to looking helpful.

But that's an anecdote. Maybe the student didn't think through the costs of her choice. How would we test what motivates people in general? We can't just *ask* them. They might lie or be self-deceived.

Kevin Simler and Robin Hanson, in their book, *The Elephant in the Brain*, propose a simple test. Consider every sphere of human activity, one-by-one. In each sphere, the institutions or activities in question purportedly aim at some noble goal, and the people involved purportedly are motivated by these noble goals. Simler and Hanson's method, then, is to ask, if people were in fact motivated by such goals, what would this predict about their behavior? If, on the other hand, they were mostly trying to look good to others, what would this predict about their behavior? They then examine people's actual behavior in that sphere.

When we examine charitable giving, we see some surprising things. First, hardly anyone does real research about whether their charities "work." They tend to trust gut feelings, or copy what others do, or do what sounds good. They respond to non-representative stories, anecdotes, and pictures rather than hard evidence and statistics. People mostly give to high-profile rather than effective charities. When givers do any research, it's usually to find evidence that whatever charity they've already chosen to donate to is good; negative evidence does not change their minds. They do not care about marginal impact; they join bandwagons instead. They exhibit confirmation bias; they ignore or reject evidence that their charity fails but accept evidence that their charity works. Only about 3% will change their donations when they confront strong evidence their chosen charities are ineffective or harmful. It might be that people simply don't trust such evidence, but it seems instead that they just aren't that interested.

It also turns out people exhibit "scope insensitivity" when it comes to charity. That means that how much good the charity claims to do has little effect on whether or how much people donate. For instance, one study finds that subjects' willingness to give to charities that claim a donation will save 2,000 birds, 20,000 birds, and 200,000 birds was about the same.[16] Some experimenters have attempted to overcome this bias through different asking techniques, but they failed.[17]

So, effectiveness doesn't induce people to give and ineffectiveness doesn't induce them to stop giving. What does? Simler and Hanson summarize the results: People give more when being watched, when primed to think about sex or dating, when they receive recognition and social benefits, or when pressured to do so by peers.[18]

If people were mostly trying to do good, we'd have to conclude they are incompetent do-gooders, stupidly throwing money away at bad charities or using money in ineffective ways. But if people who mostly want to *look* good, and only secondarily want to do good, their behavior makes sense.

If others think we are selfish, self-centered, uncaring, or uncooperative, they are less likely to befriend us, make deals with us, partner with us, or do us favors. So, we each have selfish incentives to look good to others. One way to look good is to actually be good. But we also have a selfish incentive to look better than we are. Whenever (thanks to ignorance, shortsightedness, or whatnot) there is a difference between what looks good to others and what is actually good, we have a perverse incentive to pick what looks good. This is not some novel point. Plato's *Republic* explored the philosophical implications of this very dilemma.

None of this means that we are entirely selfish. Imagine a world where everyone is a known sociopath. You wouldn't want to trick these other sociopaths into thinking you're a kind and altruistic person, because they would then target you for exploitation. Instead, the whole idea of gaming or exaggerating our reputations works only in social environments where people have a significant degree of basic altruism. They have to want to work together, cooperate, help one another, and so on.

Psychological egoism is the thesis that people regard only their own welfare as an end in itself. Psychological egoism holds that people are entirely selfish, not just mostly selfish. On this view, even when a mother feeds her crying infant, this is because the mother dislikes the sound of the cries or wants to avoid jail.

The good news is that psychological egoism is a testable theory. It's been tested and falsified.

Economists like to play games with subjects where real money is at stake, sometimes very large amounts of money equivalent to a week's or month's salary. In experimental economics games, subjects are only allowed to play if they demonstrate understanding and mastery of the rules. This ensures that the experiments are not spoiled by confused players.

For instance, the Ultimatum Game takes two anonymous players, who have their identities hidden from each other and the experimenter. One player is randomly made the "proposer" and the other the "receiver." The proposer is given, say, $100, and told to propose a split of the money with the receiver. The receiver may either approve or reject the split. If the receiver says yes, both players get the proposed split. If the receiver says no, the money vanishes, both players receive nothing, and the game ends with neither player learning the other's identity nor ever interacting with the experimenter. If people were psychological egoists, the proposer would offer $1, the bare minimum, and the receiver would accept it, since getting $1 is better than getting nothing. But, in fact, around the world, proposers usually offer more. More importantly, receivers who are offered less than 20% of the total usually reject it. This means they are willing to sacrifice their own welfare (sometimes the equivalent of a week's salary) to punish what they regard as unfair behavior from an anonymous person, even though they derive no benefit from it. Other games, such as Trust, Dictator, and the Prisoner's Dilemma, generate similar results. People are not entirely selfish.

Americans donate about half a trillion dollars per year to charity, or about 2% of their income on average.[19] They volunteered about 5.8 billion hours of work per year, with their work supposedly worth about $147 billion. Most of this money and volunteer work goes toward religious and community organizations rather than helping, say, the world's most desperate people.

I suspect that people are not only motivated by the altruistic desire to help or the selfish desire to look good. Rather, people are also motivated by principles of reciprocity and community. They want to "give back" to what they regard as the community to which they owe a debt. Pure beneficence might direct us to the neediest, but most beneficent people act on multiple principles at once.

There is another impediment to Effective Altruism. People think far more carefully when spending their own money on themselves than when spending their money on others. They think even less carefully when spending someone else's money on someone else. This is partly why we are better at buying for ourselves than buying gifts for friends, and why we tend to be wasteful spenders in democratic committees.

What would it take to get people to behave more like Effective Altruists? I ask this question in part because I don't know the answer.

Years ago, when I directed my business school's First Year Seminar, we partnered with an Effective Altruist charity. The charity gave our students a consulting problem, and 65 student teams competed to offer the best solution. The Effective Altruist charity told students that while it has an amazing bang for the buck—giving 50 cents to them does demonstrably more good than giving $100 to a typical charity—most of their funding comes from utilitarians, libertarians, and self-professed Effective Altruists. They asked students how they could attract more donors, and more ideologically diverse donors, without sacrificing their mission to provide evidence-based defenses of their work. They wanted to avoid anecdotes and what they called "poverty porn," i.e., using pictures of starving kids to elicit sympathy.

Our student groups proposed some marketing changes on the margins that helped. But none solved the problem per se. It's unclear they could. The problem is that it's unclear how Effective Altruist charities can succeed purely on Effective Altruism messaging. People don't think that way and have little incentive to change. Effective Altruist messaging is mostly not effective; it would require a large change in human nature to become effective. Having Taylor Swift talk up an Effective Altruist organization would do far more good than trying to convince the average person to become an Effective Altruist or showing people the evidence.

Notes

1 Duda 2017.
2 MacAskill 2016, 5.
3 Even these utilitarians might endorse partialism on instrumental grounds. For instance, empirically, it turns out children do better if raised by parents who love them partially. So, an impartialist

utilitarian might say that utilitarianism licenses partial commitments because they pass cost-benefit analysis.

4 Singer 1972 and Unger 1996 indeed endorse just this view.
5 https://www.roadandtrack.com/reviews/a44321399/2023-honda-accord-touring-test-drive/
6 https://www.caranddriver.com/features/g15084479/cream-of-the-crop-the-winningest-cars-in-10best-history/#
7 MacAskill, Todd, and Wiblin 2015; Aos et al. 2004; Petrosino et al. 2013.
8 https://betonit.substack.com/p/you-dont-have-to-be-a-moral-saint
9 Schmidtz 2006.
10 Oakton High School, where I help, is certainly not the neediest school, not even neediest in terms of musical talent. On the contrary, they are overflowing with talent, if not *guitar* talent. But the point of helping them is to build community, not to meet needs.
11 A few years ago, I had a student with just such a dilemma. She was offered an analyst position at Bain and a secretary position at a some small non-profit. I gave her the standard Effective Altruist advice. The next best secretary was likely as good as she is (indeed, this wasn't a good use of her talents) and she could pay for three or four secretaries with a portion of her yearly bonus. Working at Bain would put her in a position to enable more people to do the kind of good charities do.
12 Ross 1930, 21.
13 It can be distinguished from civic duty, which is the obligation to provide for the good specifically of one's community, and from special obligations, like the duty to provide for the good of one's friends and family.
14 For a comprehensive review of the empirical benefits of wealth in terms of health, culture, cooperation, and so on, see Brennan 2020, Chapter 2.
15 Schmidtz 2006, 91.
16 Desvousges et al. 2010.
17 Maier et al. 2023.
18 Simler and Hanson 2018.
19 National Philanthropic Trust 2023.

7
ARNOLD'S RESPONSE TO BRENNAN

Effective Altruists hold that we should try to improve the world, and that our efforts to do this should be guided by reason and evidence. They think we should *do good, better*.

This idea is at once simple and surprisingly slippery. It's hard to pin down just what Effective Altruism entails—and what it doesn't. Is it merely applied utilitarianism? How demanding is it? Does it require lots of financial and personal sacrifice? What's its relationship to philanthropy?

Jason Brennan's entries do much to clarify and illuminate Effective Altruism. Effective Altruism, Brennan explains, is not just "rebranded utilitarianism." Instead, it's "a set of substantive intellectual tools meant to help us transform our benevolence into beneficence." We can use these tools to "ensure that when we try to help others, we succeed."

Effective Altruism, Brennan convincingly shows, is compatible with a wide range of philosophical views about the nature and strength of our obligations to aid others. While some utilitarians, like Peter Singer, posit very demanding theories of required aid (roughly: give until further giving would hurt you nearly as much as it helps others), one needn't accept these theories to be a card-carrying Effective Altruist. Suppose you hold the commonsense view that people should indeed aid others—but only sometimes, and not in any super-demanding way. "Even people with such wishy-washy views," writes Brennan, "can endorse Effective Altruism." This is because even this weak-tea "imperfect duty" to help others is still a duty to *help*: not a duty to *look* like you're helping, or to *want* to help, but a duty *actually to improve others' well-being*. But to discharge *that* duty, one needs precisely the sorts of tools provided by Effective Altruism.

Indeed, even someone who denies that there's any duty to aid others at all can be an Effective Altruist. Take Ryan Davis. In his contributions to this volume, Davis argues that beneficence is morally optional. It's nice to do it, but not wrong if you don't. Well, even Davis can be an Effective Altruist, on Brennan's view, in the sense that even Davis has reason to welcome tools that "ensure that when [he tries] to help others, [he succeeds]."

Brennan also persuasively argues that Effective Altruism is compatible with different ways of helping others. Effective Altruists should be operationally flexible; they shouldn't fixate on any particular method for doing good, like philanthropy, or "earning to give," or volunteering. Instead, they should adopt whichever methods work best. As Brennan explains:

We shouldn't fetishize anti-malaria pills over mosquito nets. We should pick the intervention that actually works. But this generalizes. What matters is what's effective, not whether helping comes in the form of volunteering, charity, government, social business, or regular old business.

Business? As a vehicle for beneficence? Yes, indeed. Brennan contends that business (whether "social" or "regular") can be an effective—and indeed, even superior—alternative to charity. Key here is what he calls the "central philosophical problem of charity," namely, that in the "absence of a profit-loss accounting mechanism, it is difficult to judge whether a given charity is creating or destroying value." Is the charity actually helping people, or might it instead be harming them? Without the feedback provided by market signals, it's hard to tell.

Businesses, Brennan argues, don't face this same difficulty. *Assuming certain background conditions obtain* (a point to which I'll return), they can rest assured that if they're making money, they're creating value; if they're not, they're not. Importantly, this logic holds not just for so-called "social enterprises"—businesses that "use for-profit market mechanisms to supply the goods or services traditionally supplied by charity or government"—but also plain-old, vanilla businesses that are just trying to make money.

In one of his entries' most striking passages, Brennan writes that "the typical person, working the typical job at the typical employer, already acts beneficently by *doing that job*. They already discharge at least some of their obligations of beneficence through such work." Further,

> most of us do far more for others through our productive work than through charity, not merely because we are bad at selecting effective charities ... but because regular work does far more good for others than it gets credit for.

Another unexpected tool in the Effective Altruist toolkit, according to Brennan, is consumption. It's not just ordinary working stiffs who contribute to the general good; it's also the much-maligned consumers who buy what those stiffs are making and selling. As former NYC Rudy Giuliani (admittedly, not a noted moral philosopher) put this point in the wake of the 9/11 attacks: "There is a way that everybody can help us. Come here and spend money. Like, go to a store, do your Christmas or holiday shopping now, this weekend."

Brennan notes that the twentieth century's astounding increases in prosperity—and thus, in well-being—were caused by *commerce*, not charity. "It's not as though Japan and Korea became rich because people followed, say, Peter Singer's advice, and donated 75% of their incomes," writes Brennan. "Instead, they became rich precisely because people ignored Singer's advice." Capitalists invested; laborers labored; and consumers consumed. And "hundreds of millions of people were liberated from poverty."

Given my publication history, one might think I'd disagree with much of Brennan's account. In the past, I've written favorably about socialism,[1] negatively about capitalism,[2] and critiqued consumerism.[3] Furthermore, in my third entry for the present volume, I argued that businesses should largely forbear from socio-political activism. And yet, despite all the shade I've thrown capitalism's way over the years, I find myself largely persuaded by Brennan's argument that ordinary commercial activity—working, buying, and selling—can, under the right conditions, redound powerfully to the common good.[4] It therefore deserves a central place in the Effective Altruist's toolkit.

But I do have questions for Brennan.

First question. What should Effective Altruists make of "corporate social responsibility" (CSR) and "Environmental, Social, and Governance (ESG) investing"? CSR says that businesses should aim to be "socially responsible," meaning, they should aim to benefit a broad group of "stakeholders"—a group that includes not just shareholders, but also workers, suppliers, customers, and community members. Socially responsible businesses often hew closely to ESG criteria, which purport to measure a company's "triple bottom line"—its impact on planet, people, and profit.

Brennan has argued persuasively that (with some background assumptions in place) regular old commerce benefits humanity. Does CSR/ESG-inflected commerce benefit humanity even more? Proponents would surely say that it does. After all, socially responsible corporations *explicitly aim* to promote the good even as they make a buck. However, explicitly aiming at an outcome isn't always the best way to promote that outcome. Sometimes, an outcome is best promoted indirectly, as a byproduct of some other intentional action. (See Adam Smith's "invisible hand.")

Does business produce more value—hence, act more beneficently—when it sticks just to maximizing profits? Granted, proponents of CSR and ESG often tout the "business case" for the stakeholder approach; they claim that socially responsible businesses are just as profitable as their regular counterparts. But surely that's not *always* so. Consider a case where profit-maximization and social responsibility/ESG criteria come apart. How should we think about such cases from the EA perspective?

To package this first question slightly differently: Milton Friedman famously said that the social responsibility of business was to maximize profits. In light of Brennan's argument, could we add that, by maximizing profits, Friedman-style businesses further the social good as well as any so-called "socially responsible" business? (And if so, what's the point of CSR or ESG?)

Second question. Brennan is careful to emphasize that businesses work their beneficent magic *only when* markets are "competitive, free, and unsubsidized," and when there are "no significant externalities." Absent these conditions, profit no longer reliably signals value creation. For example, when government subsidies artificially inflate demand for some good, businesses can make a killing supplying that good—even though doing so contributes inefficiently to well-being. Or again, if market transactions impose significant costs on third parties—i.e, if they create "negative externalities"—then profitable transactions may be value-destroying on net. So, in making the EA case for business, we've got to remember that there's a big asterisk attached: business activity reliably creates value *only when* certain background conditions are met.

The trouble is, these background conditions frequently *aren't* met. In the real world, many markets are uncompetitive; governments subsidize all sorts of things; and market transactions impose negative externalities galore. So, Brennan's defense of businesses' EA bona fides is less powerful than one might have thought. To what extent does Brennan's argument vindicate *actual* businesses operating under *realistic* conditions?

Let me linger on the problem of negative externalities, because it poses an especially thorny problem for Brennan's view. Market transactions systematically fail to internalize costs falling on two groups of great interest to many Effective Altruists: animals and future generations.

Tyson Foods—one of the world's largest producers of meat and poultry—reported a gross profit of $6.6 billion in 2022. You might think: "Wow. That's a lot of value. How beneficent." But in fact Tyson Foods is probably one of the worst companies in the world, from an EA perspective—at least, if we include the well-being of all sentient creatures in our calculations, as we surely should. Tyson made a lot of money, and created a lot of happy humans, but at what cost to animal welfare?

The point generalizes far beyond Tyson Foods, of course. A great deal of business activity imposes horrific "neighborhood effects" or "negative externalities" on animals. I see no easy way to internalize these costs. I'd be curious to hear Brennan's thoughts on this issue.

Another reason to doubt the "profits signal beneficence" thesis is that much economic activity imposes negative externalities on future generations, who, like animals, have no way to push back. In 2022, Exxon Mobil Corporation reported a net profit of some $56 billion. Again, this might *seem* to indicate lots of EA-approved, beneficence-boosting activity. But of course, much of this profit comes at the expense of future generations, on whom Exxon (and its consumers; our hands are not clean here) shunt the potentially catastrophic costs of climate change. *I* benefitted from buying and burning a bunch of gasoline this year; Exxon's products certainly contributed to *my* well-being, and, I'm sure, yours as well. But how will they affect people in 2040, or 2100?

Again, I don't mean to pick on any one company in particular; nor do I take myself to be raising a narrow or niche point. Since *nearly all economic activity* involves the use of fossil fuels; and since said usage imposes steep negative externalities on future generations; how reliably does *any* company's profit indicate a net gain in well-being?

Turning to a third and final question: let's talk about Brennan's genie. This beneficent fellow offers you three choices:

A. "He donates $100 to charity. The charity does $120 of good for others.
B. He volunteers for one hour at a charity. The cost to him of his labor is $100. His work does $120 of good for others.
C. He works for one hour at a factory. The cost to him of his labor is $100. His work does $150 of good for others."

"If these are your only choices," Brennan writes, "you should choose C ... Otherwise, you are fetishizing how you help over how much you help."

Does it matter *who* you help? I know that many Effective Altruists, being utilitarians, say that well-being is well-being, and it doesn't matter which specific container it goes in. You should be impartial across people. 10 hedons is 10 hedons, whether they go to Elon Musk or a homeless person. But from a different, more "prioritarian" perspective, it matters very much whom your activity benefits. Better to benefit someone worse off, less, than someone better off, somewhat more. Or so say prioritarians.

If that's right, then an EA would need to know more about options A, B and C before choosing between them. If the charity produces $120 of value by, say, deworming African children, whereas the factory produces $150 of value by, say, producing lululemon tights for wealthy yoga moms, I'm inclined to say: A or B it is.

Notes

1 Arnold 2016; 2022a.
2 Arnold 2017; 2022b.
3 Arnold 2021.
4 Does this mean I recant my former work? No, because none of it conflicts with Brennan's claim that business and commerce promote well-being. My objections to capitalism and consumerism hinge on other values like equality, freedom, community, and democracy. I grant that capitalism delivers the goods, and thus increases well-being (at least, if we factor out "negative externalities" falling on animals and future generations—a point I develop later in this piece). I worry, though, that capitalism increases well-being only at serious cost to other important values. This is not to say that socialism would be better in practice; I'm pretty sure it wouldn't.

8
CHAPPELL'S RESPONSE TO BRENNAN

Question 4: Can We Endorse Effective Altruism Without Endorsing Utilitarianism?

The two most distinctive and controversial features of utilitarianism are:

1. its rejection of intrinsic deontic *constraints* against harming some as a means to more greatly benefiting others, and
2. its rejection of personal *prerogatives* to do less than the best (when the best act would be personally costly to you).

Effective Altruism is not committed to either of these controversial features. So you can endorse Effective Altruism without endorsing utilitarianism. The philosophy here is straightforward. What's more interesting is the sociology. Why do so many philosophers seem tempted by the obviously false claim that Effective Altruism is just "rebranded utilitarianism"? And why does one find comparatively few non-utilitarians actually endorsing Effective Altruism?

One possible answer is that there's more to a moral theory than just its formal features. At a deeper level, a moral theory represents a kind of moral perspective on the world. Utilitarianism represents a moral perspective on which beneficence is *exhaustive* of what matters. Effective Altruism represents a moral perspective on which beneficence has *central* importance. There's obvious overlap between the two, even if Effective Altruism, unlike utilitarianism, is compatible with endorsing other fundamental values in addition to beneficence.

I like the term 'Beneficentrism' as a label for the implicit moral principle that motivates many Effective Altruists: that we all should embrace a non-trivial commitment to (efficiently) advance beneficent ends, because everyone's well-being matters immensely.

Beneficentrism is compatible with pluralistic moral theories such as commonsense deontology. Our beneficent efforts may be constrained by respect for human rights or other moral constraints. And beneficence needn't take over our lives: you can be very beneficent (with, say, 10% of your resources) while also pursuing other personal or family goals, from parenthood to pet projects.

So why don't more deontologists embrace beneficentrism (and, by extension, Effective Altruism)? I wish I knew. It doesn't seem to me that there's any *good* reason for their reticence. But I can think of three (bad) possible contributing factors:

1. Some deontologists are drawn to the view that "numbers don't count": we have no more reason to save a million people's lives than to save just one distinct individual.[1] But most people (rightly) find it obvious that it's better to save more lives, all else equal, so let's set aside the anti-numbers view as a non-starter.
2. I suspect a larger factor is that non-utilitarian theories simply don't make the importance of beneficence as *salient* as utilitarianism does. Many moral philosophers are fixated on identifying the dividing line between *permissibility* and *impermissibility*. Since much beneficence is intuitively "above and beyond the call of duty"—or what philosophers call *supererogatory*—they give it short shrift. It isn't where their attention is focused. But this is arguably a mistake.[2] Saving and improving lives should be regarded as highly *important* on any reasonable moral view, no matter whether it's ultimately classified as obligatory or supererogatory.

   As moral agents, we should not set our sights so low as to merely do the *bare minimum* required to avoid outright impermissibility or blameworthiness. We should instead try to act *well*, and be guided by consideration of what's important. Finding that it's *permissible* to be a selfish jerk wouldn't

settle that it's *the thing to do*. Once more deontologists take this simple lesson on board, it would seem harder for them to avoid the conclusion that it's worth embracing beneficentrism after all.
3. The failure of past deontologists to take beneficence sufficiently seriously may also mean that the most naturally altruistic people will find the utilitarian moral tradition more congenial to their values than its competitors. There's no logical reason why this must be the case. But psychologically, if utilitarians are almost the only philosophers around who seem to take beneficence seriously, then it's understandable that beneficentrists would gravitate in their direction. We're drawn to the moral theories that best represent our moral perspective on the world, and the visions presented so far by actually-existing (capital-D) Deontologists haven't tended to give much weight to reasons of beneficence—even if there's nothing in the underlying theory of deontology that would justify this neglect.

Hopefully that will change in future, and more beneficentrists who are suspicious of the utilitarian rejection of constraints and prerogatives will find a friendly home among beneficentric deontologists—who agree with a large proportion of utilitarian claims, but (crucially) not all.[3]

Question 5: Is Beneficence Confined to Charity and Volunteering, or Can We Exercise It in Other Domains, Including Business?

I wholeheartedly agree with Brennan that we can do good via all sorts of means, including many ordinary jobs and marketplace exchanges, even in the absence of any altruistic motivation. That said, it would be an incredibly convenient coincidence if our self-interest perfectly aligned with our reasons of beneficence. So I also agree that we can't just assume that ordinary consumer spending is the best that we can do. I'll further argue that past moral convenience may increase what can subsequently be asked of us.

First, a worry: Brennan's invocation of East Asian development could be misleading. The growth of these economies since 1950 is indeed one of the great success stories of human history. We should hope for more developing countries to follow in their footsteps. And the potential for global trade to alleviate poverty in this way provides strong reasons to oppose illiberal economic policies that could slow or stall such development.

But the suggestion that poverty was alleviated "precisely *because* people ignored Singer's advice [to donate money instead of purchasing luxuries]" implies that less good *would* have been achieved had Westerners instead directed all their excess wealth to effective charities. And there seems no basis for that counterfactual claim.

Brennan notes that "we have no record of charity solving the problem of extreme need." But total donations to effective charities are measured in the mere *millions* of dollars annually.[4] (And before GiveWell was founded in 2007, there was basically no rigorous evaluation of charity effectiveness.) Global trade, by contrast, is measured in the *trillions* of dollars each year. We have no record of *anything*—including trade—"solving the problem of extreme need" with mere millions of dollars. Effective charities have nonetheless achieved a remarkable amount, including the eradication of smallpox—which alone amounts to millions of deaths averted *every year*.[5] If our society were to deliberately direct *trillions* of dollars annually to the very best causes we could find, it beggars belief to think that this could not achieve even more good than spending all that money on consumer electronics.[6]

There are tradeoffs to consider between near-term and longer-term impacts. If investing in research & development would save more future lives than funding famine-relief today, then I would prioritize the former. But not all investments are equal. Again, it would be a remarkable coincidence if investments intended to optimize for profit (in an unequal world) were also the most socially beneficial investments. There's more profit to be made in curing male pattern baldness than malaria, after all, since the wealthy are vastly more likely to suffer from the former than the latter. But we should presumably expect the latter to do more long-term good.

Brennan offers some systematic reasons to be skeptical of charities, including the epistemic difficulty of measuring value and the misaligned incentives that stem from the intended beneficiaries not being the ones who hold the purse strings. Brennan's solution is "social businesses" that cater to the world's needy. But the fundamental problem of poverty is precisely that the poor cannot afford to purchase all that they need (let alone all that would be of genuine *value* to them). Treating payment as the measure of value, while leaving existing inequalities untouched, blatantly undervalues the interests of the poor.

A crucial prior step, I would think, is redistribution of wealth via direct cash transfers (as offered by the charity *GiveDirectly*), so that the global poor have the resources to buy what they value. There's no epistemic difficulty to the question of whether transferring wealth from the rich to the poor adds or destroys value.[7] We all know that $100 is worth more to the poor than to the rich (including the average citizens of wealthy nations like the United States). And there are no misaligned incentives either: GiveDirectly beneficiaries end up holding the purse strings, and can make informed purchasing decisions based upon their personal needs and desires.

That said, we shouldn't exaggerate the problems Brennan raises for other charities. *GiveWell* seems confident that they're able to reliably identify charities that are many times more cost-effective than cash transfers—their "bar" for recommending funding, as of September 2023, is cost-effectiveness in excess of "10x cash"[8]— so the difficulties must not be insurmountable. But I'd certainly recommend cash transfers over an *average* or randomly selected charity, partly for the reasons Brennan points out.

Finally, I want to consider the significance of Brennan's observation that "While earning to give can be noble, most of us already *give by earning*." I'm very open to the suggestion that "regular work does far more good for others than it gets credit for." But how far that goes toward discharging our duties of beneficence depends on how much better we could do.

I've elsewhere defended the view that we ought to (at least) do the most good we can *without* suffering undue burden.[9] I formulate the relevant sense of "burden" in terms of an individualized effort ceiling (how much effort it would take you to be a decent person, roughly speaking), but you could stick to traditional welfare costs if you prefer.

This moral principle allows us to secure two plausible thoughts about *unintentional beneficence*:

1. It's better to do more good unintentionally rather than less good intentionally. You should never deliberately choose or prefer to do less good, merely to make it more intentional.
2. Good side-effects from doing what you self-interestedly prefer don't get you off the hook for doing *more* good when the opportunity arises.

We secure the first thought because, between the options described, the greater good can be achieved at no cost to yourself. So it is the most good you can do (between those options) without undue burden.

We secure the second thought because you haven't yet taken on any moral burden *at all*. So you can hardly complain if morality asks you to do a bit more. "*Look how much good I already did while simply pursuing my self-interest!*" is not a compelling excuse to refrain from trying a little in order to do even *better*. It's excusable to do less than the best when the best would be an undue burden. But having previously done good inadvertently is not the mark of suffering under an undue burden. Quite the opposite.

So, yes, a wide range of (even ordinary) actions can have beneficent effects, and be well worth doing. But benevolent motives may still be relevant by helping to determine *how much* beneficence may be demanded of us, since it's hardly unreasonable to ask for *some* minimal moral effort (to secure even better results) on top of whatever you can achieve with no moral effort at all. The low-hanging fruit of *convenient* beneficence is good as far as it goes. But however far it goes, this gives us *no* reason, and no excuse, not to do *even better* when the opportunity arises.

Question 6: Why Are People So Bad at Charity and What Should We Do About It?

People tend to have a complex mix of values and motivations, and which ones get activated in practice can depend a lot on contextual factors (such as *salience*) that lack objective moral significance.

For example, if we see a child drowning in a shallow pond right before our eyes, most of us would feel highly motivated to save them, even at some financial cost to ourselves (perhaps we're wearing an expensive suit that will be ruined). If we merely hear, abstractly, that there is a child far away whom we could save at equal cost, we're apt to feel much less motivated to help. There's no objectively morally relevant difference between the two cases:[10] being out of our line of sight doesn't make the distant child *matter* any less. But it does make their interests less subjectively *salient* to us, so our sympathy and other moral emotions are less likely to be activated in the ways that are objectively warranted.[11]

A key problem of moral motivation is that we struggle to process purely abstract information appropriately. Emotional activation is often essential (or at least very helpful) for moral motivation, and abstract information isn't well suited to rousing an emotional response.

In light of this fact, we often *need* more vivid forms of communication—including pictures and anecdotes with gripping narratives—in order to process information appropriately. If a charitable organization refuses to appeal to our moral emotions during fundraising, this amounts to a failure of communication on their part. They *should* be trying to rouse our sympathy: that's precisely what we need to be able to process the information appropriately. Presenting ordinary people with raw numerical data is about as helpful as presenting a graph to a blind person. Accommodations need to be made that take into account the reality of our cognitive and motivational limitations.

Still, even after someone's altruistic motivations are activated, it's a further step to get them to prefer *more effective* ways of helping over less effective ones. In "The Psychology of (In)Effective Altruism," psychologists Lucius Caviola, Stefan Schubert, and Joshua Greene survey a range of cognitive and motivational obstacles to effective giving, and suggest possible solutions.[12] One solution they have successfully implemented—see givingmultiplier.org—is to reduce the cognitive burden of tradeoffs by encouraging non-EA donors to *split* their donations between their favorite charity and a more effective one. It's not as ideal as 100% effective giving, but it gets a lot more people willing to give effectively at all.

Another thing that could help is shifting social norms. Many Effective Altruist organizations, such as *Giving What We Can*, encourage people to be more open about the value that they place on effective giving. Part of the hope is that this may help to "normalize" effective giving, and encourage more others to act similarly. We're often influenced by those around us, so it's not unreasonable to hope that growing numbers of Effective Altruists could (over time) shift our social norms in favor of more effective philanthropy.

Failing that, it may just be that people with certain personality types, cognitive styles, and/or educational backgrounds are disproportionately more likely than others to be receptive to the idea of effective altruism. If others continue to find it emotionally unappealing (no matter the strength of the arguments in its favor), this may place a low ceiling on the Effective Altruism movement's capacity for further growth. That would be a sad result. But it shouldn't stop those who *are* drawn to effective altruism from continuing to do good effectively—including by appealing to non-EAs in ways that the latter are more likely to find motivating.

Notes

1 Taurek 1977. In response, see Parfit 1978.
2 I argue this in greater detail in a new paper I'm working on, titled "Permissibility Is Overrated."
3 For example, Pummer 2023.
4 See, e.g., https://www.givewell.org/about/impact: before 2015, the most "money moved" by GiveWell recommendations to effective charities in a single year was $35 million.
5 MacAskill 2016, 46.
6 For more on what could be achieved with just $3.5 trillion, see the Longview Philanthropy report, *What if the 1% gave 10%?*, https://longview.org/what-if-the-1-gave-10/.
7 At least when only considering direct effects, as is also the case in Brennan's profit-based accounting. A full moral accounting would need to consider the risk of unpriced negative externalities, such as the harm to non-human animals from increased meat consumption.
8 https://www.givewell.org/how-we-work/our-criteria/cost-effectiveness/cost-effectiveness-models (accessed October 6, 2023).
9 Chappell 2019a.
10 Peter Singer 1972 and 2009 famously defends this verdict.
11 Chappell and Yetter-Chappell 2016.
12 Caviola, Schubert, and Greene 2021.

9
DAVIS'S RESPONSE TO BRENNAN

Jason Brennan's essay explores how Effective Altruism can work in real life. He wants to think about people as they are, trying to understand how we could do better by starting off with a realistic look at why we aren't doing very well.

I'm persuaded by much of what Brennan says. I share the basic idea that a broadly social scientific understanding of humans complicates thinking about how they might be encouraged to act more beneficently. But as I read Brennan, there's good news as well as bad news. The bad news is that it's not easy to push humans to be more beneficent. The good news is that our assessment of others' beneficence is biased in many of the same ways that their beneficence is biased. Just as people want to signal they *are* good more than they want to *do* good, we are tempted to assess others based on their signals more than on how much good they're doing. So we judge them on what they're sacrificing, not on how much they're helping. As Brennan points out, that makes no sense. Once we think clearly about what actually matters, many self-interested or even apparently selfish actions might count as beneficent. And that means even selfish people might be acting beneficently. So perhaps the glass is half-full, after all.

My questions in this reply will be less objections, and more inquiries about how to continue to think about these issues.

How Does Beneficence Compare with Other Values?

Brennan understands Effective Altruism as "a set of substantive intellectual tools meant to help us transform our benevolence into beneficence." These tools might be congenial to utilitarians, but one need not be a utilitarian to use them. You could also be a Kantian or an Aristotelian (whatever they might believe). I think there is an insight here worth drawing out just a little more. The utilitarian supports Effective Altruism as the best way of promoting something like overall well-being. The Kantian doesn't care about well-being but does care about respecting the rational agency of persons. Many Kantians believe that maintaining or promoting the rational capacities of other persons respects their rational agency.[1] So, Kantians might support Effective Altruism as the best way of maintaining the rational capacities of other persons.

What's interesting about this is that for the Effective Altruist's purposes, it doesn't really matter which background theoretical outlook you happen to favor. It's not as if there's one altruistic agenda that focuses on the well-being of the globally worst off, and another, separate agenda that seeks to help maintain the rational capacities of the globally worst off. That's because conditions are bad enough that the threats faced by the most vulnerable populations affect their well-being and capacities alike. Extreme poverty and debilitating disease are bad from utilitarian and Kantian perspectives.

This might make it seem like from an epistemic standpoint, things are straightforward. We don't need to resolve thorny philosophical matters to know what we morally ought to do. But this is not what Brennan thinks. Instead, he suggests that reasons of beneficence are one kind of moral reason among at least several: reasons of reciprocity, and reasons of community, to name a couple. Brennan observes that "in the real world, other value considerations are often at play, and it's often not obvious what does the most good."

My question is about how we should interpret this conclusion. One idea is that we should be epistemically uncertain about what will do the most good, impartially considered. A second idea is based on a kind of pluralism about values. The impartial good is one source of moral reasons, but it's not clear how that relates to what we ought to do, all things considered. Maybe our reasons of community and reciprocity are sometimes more important. A third consideration is that we enjoy a kind of moral prerogative. Perhaps there is no all-things-considered ought, or at least no all-things-considered requirement, about how to weight differing moral reasons.[2] So when you are in a position in which beneficence points one way and reciprocity or community points another, you enjoy some discretion about what to do. Of course, there is no tension in these explanations. My question is: What is the sense (or senses) in which it is not obvious what does the most good? Is this about our epistemic limits, or about the metaphysics of value, or about our agency in determining our own reasons, or perhaps something else entirely?

How Much Good Are We Obligated to Do?

Brennan's next section is about what he calls the *central management problem of charity*. The problem is that charities' customers are not their beneficiaries. This is a problem because institutions are motivated to do what their customers want, so there's a mismatch between what charities are motivated to do (promote their customers' values) and what they're supposed to do (promote the good of the recipients of their efforts). Brennan suggests that we should expand our idea of beneficence to include whatever does the most good for others. Maybe doing one's job does more for others than one's charity work, or even than donating money. The point is not to make an empirical claim, but to raise a question. Once we notice that we should simply care about what helps other people, it might well turn out that what is best for others is also something that is good for us as well. As Brennan points out, you should not "fetish[ize] how you help over how much you help."

I want to ask about whether this claim has any relation to the question of how much we're obligated to help. Consider Brennan's case of the worker who takes a job through which they make $2 million over the course of their career, and contribute some value greater than $2 million to others. Let's call this person the Successful Careerist. Now imagine someone else—maybe Successful Careerist's college roommate—who has the exact same opportunity to get a job through which they would contribute a comparable value to others. However, in their last year in college, they decide that what they really want to do is live off the grid in a cabin in British Columbia, maintaining a small sustainable garden and cutting firewood to get by. College Roommate's choice of a simple life means that they forgo providing a great benefit to others through a productive career.

My question is: Is there anything morally amiss about this?[3] My own view is that it's completely permissible. As Mary Oliver and Ralph Waldo Emerson agreed when they imagined cases like this, you're under no demand to live your life for other people. My guess is that Brennan will agree, too. My textual evidence for this guess is his description of competing values, as described above. Next, assuming I'm right in my guess, I want to ask about the Successful Careerist. Is it permissible for Successful Careerist to keep their money, rather than to become an Effective Altruist and give it away?

Just to spell the puzzle out a little more: If we answer yes, it is permissible, then it seems like we've reached a substantive and perhaps surprising conclusion. That is, you are not morally required to give away the money you make in your successful career (though it would be great if you do). On the other hand, suppose the answer is no, it is not permissible to keep their money. If that is true, and if we agree that College Roommate can permissibly move off the grid, then the consequence is that someone can live their life in a way that promotes value for others to the tune of >$2million, and yet still be acting morally worse (in terms of beneficence) than someone who contributes no value to others at all. That seems like a surprising outcome.

In short, my question is about what inferences, if any, Brennan draws from his observations about work and beneficence for the degree to which beneficence is morally required.

Which Moral Beliefs Can We Trust?

An upshot of Brennan's essay is to complicate the question of how we can best help others. He finds it "far from obvious what the optimal mix of trade, investment, and giving is." I want to ask whether Brennan's empirical discussion might also add a further complication. If we understand humans—as he does—as apt to be self-deceived and motivated in their moral reasoning, then what should we think about guiding our actions based on human moral intuitions (even including our own)?

Let's recap the evidence Brennan cites. Most of us are highly motivated by a kind of social desirability bias. We really want others to see us as good, even more than we want to be good. And we can be taken in by our own act, just as much as anyone else can. In fact, it's often important that we are taken in by our act, because that's how we can best sell the image of ourselves as morally motivated and altruistic. A theme in (my reading of) Brennan's earlier work is that these biases can also afflict philosophers as much as anyone else.[4] For example, many political philosophers are so committed to a certain schedule of democratic virtues that they won't abide empirical evidence to the contrary.[5] In general, political psychologists find that cognitive sophistication does not eliminate bias, and might even make it worse.[6]

My question is about what confidence we should have in common moral intuitions, given this predicament. If most people are acting in ways designed to look good rather than to be good, shouldn't we also expect that the moral attitudes they express represent what-it-would-be-socially-desirable-to-say, rather than the product of some truth-detecting faculty of moral reasoning? In other words, if we don't trust people when it comes to their moral actions, why trust them when it comes to their moral intuitions?[7]

You might think that philosophers can somehow escape this problem, because perhaps their intuitions about morality have received some disciplinary training. The evidence is against this hypothesis.[8]

I'm raising this question because if Brennan agrees that we should be skeptical of humans' intuitive moral judgments for the same kinds of reasons that he recommends skepticism about their putatively moral actions, then where does this leave philosophical inquiry? One possibility is to accept a theoretical outlook that's only minimally (if at all) connected to moral intuition. This is the kind of position I attribute to Chappell in this volume. However, unlike Chappell, Brennan starts his essay by distancing himself from any theoretical commitments about morality. I agree nothing like that is needed to be an Effective Altruist. But if we accept the kinds of skepticism about self-understanding that might motivate the Effective Altruist elsewhere, what kind of moral epistemology are we left with? To what sources should we look for moral guidance?

Conclusion

The question Brennan ends with is one he says he doesn't know the answer to: How could we get people to behave like Effective Altruists? I haven't said anything helpful about that, but I have agreed that the way to start is to think about how people are motivated to act. My questions have centered on how real-world moral motivations might bear on our beliefs about normative ethics and moral epistemology. But set aside those worries for now. I want to end by pointing out that while Brennan says he doesn't know the answer to his final question, his last sentence proposes a pretty good idea: Maybe we could get Taylor Swift to help?

Notes

1 Smith 2011; Korsgaard 2009; Herman 2001.
2 For a relevant discussion, see McPherson 2018.
3 See, for discussion, Buss 2006.
4 Philosophers have strong opinions that seem sometimes at odds with empirical work on the topic. See Brennan 2014; Brennan and Magness 2019.
5 Brennan 2016.
6 Rekker and Harteveld 2022; Vegetti and Mancosu 2020; Flynn, Nyhan, and Reifler 2017.
7 Machery 2019.
8 Maćkiewicz, Kuś, and Hensel 2023.

10
BRENNAN'S RESPONSE TO ARNOLD, CHAPPELL, AND DAVIS

Response to Arnold

Arnold asks how Effective Altruists should think about stakeholder theory, ESG, CSR, "triple bottom line," and similar business strategies. My answer: they should probably ignore them.

Every few years, academic business ethicists or corporate leaders produce some new "theory" of business ethics, which usually gets some acronym as a label. The supposed goal of such "theories" is to help business leaders make more ethical decisions, both by giving them a normative standard and by providing a decision-procedure or heuristic for good decision-making.

I put "theory" and "theories" in scare quotes here because I worry stakeholder "theory," ESG, CSR, the triple bottom line, and related ideas are neither robust nor substantive enough to constitute theories. To say they are bad theories would give them too much credit; they don't have enough meat on their bones to qualify as rotting meat. To illustrate my worry, imagine I offered a new "theory" of business ethics, as follows:

The RARIAC Model of Business Ethics

When a business makes a decision, it should properly respect the rights and weigh the interests of all those with a legitimate claim against it, and should properly consider all morally significant consequences of its actions.

The problem with my new "theory" is not that it's false; it's that it's trivially true. It doesn't say much of anything.

Every moral and political outlook, every moral and political theory, and every normative stance says that agents should properly respect all rights, properly weigh all legitimate interests, and properly consider all morally significant consequences. (Even moral nihilists or ethical egoists say that.[1]) What substantive theories debate is which interests are legitimate, which rights people have, what counts as proper weighing or respect, which consequences matter and how, what it means to properly consider all morally significant consequences, what tradeoffs we may make, which side constraints we must respect, and so on. The sentence above says little more than "Business ethics means doing the right thing." It's not a theory, but a platitude.

Suppose I want to generate some paid consulting gigs as a business ethicist. I need to generate some acronym for my platitudinous non-theory. So, let's call it RARIAC—Respect All Rights, Interests, and Consequences. I could then make arguments like this:

The Missing Premise Argument

1. Businesses should always abide by RARIAC. [A vacuous claim that can't possibly be false.]
2. Therefore, business should donate 10% of their profits to [some sexy NGO my colleagues and I like] and [do some left-wing management stuff we also like].

In poking fun at such things, I'm not even disputing whether businesses should donate to various NGOs or use various left-wing management practices. My point here is that you can't derive substantive claims like 2 from vacuous claims like 1.[2] You need intermediary principles with more substance.

When you examine academic articles or business white papers on CSR, ESG, the triple bottom line, stakeholder theory, and so on, you find many arguments like this. These papers start with vacuous and trivial statements, such as that businesses should properly weigh and consider the impact of their choices on the environment, society, and governance. (Of course, they should!) They then make particular recommendations *out of nowhere*—such as that we should recycle more or hire more minorities. They don't supply or defend the missing intermediary principles needed to generate these substantive conclusions. The pattern is:

platitudinous slogan → controversial conclusion.[3]

Thus, when Arnold asks whether ESG, CSR, stakeholder theory, and like can be forms of Effective Altruism, my answer is *of course they can*. They can also be the opposite. They don't really say anything, so they can be anything.

ESG, CSR, stakeholder theory, and the like are intellectually vacuous. However, they might have some value despite their vacuity. An interesting and persisting finding in moral psychology is that people often act wrongly not because they consciously choose to do the wrong thing, but because they operate on auto-pilot and fail to notice when they have entered morally charged situations.[4] Because of this, sometimes training people to ask broad questions, such as "Whose interests are affected?" induces them to notice morally important factors and then act better, even when these people have no training on how to *answer* these questions.

That's not because "properly respect all legitimate interests" is a good theory—it's platitudinous. It's just that paying attention turns out to be better than not paying attention. Stakeholder theory, ESG, the triple bottom line, and so on, are less like theories which answer hard questions and more like lists of things to consider. ESG, the triple bottom line, CSR, and the like might be helpful for overcoming our moral blind spots, despite my criticisms.

But that doesn't mean these platitudinous "theories" are overall useful or good for business leaders. Sometimes corporate leaders use such "theories" as excuses or to disguise their business failings—a leader asserts that he failed to turn a profit (something easy to measure) because he instead promoted some impossible-to-measure, high-falutin' moral goal. (If I say I was following ESG, how could you prove me wrong? I can give an ESG justification for almost anything I do.) Sometimes, worse, corporate leaders use ESG and the like to greenwash or ethics-wash their bad business practices. They don't act better, but they learn to talk as if they act better. It's thus an open, largely empirical question whether getting business leaders to think explicitly about ESG, CSR, and the like makes them behave better.[5]

In contrast, I view EA as a substantive theory with particular methods, such as applying basic principles of economics and behavioral psychology to our giving behavior. However, EA is not a theory of business ethics, because the scope of business ethics is distinct from what EA covers. For instance, an important question in business ethics is what constitutes improper deception in marketing. EA takes no stance; it's not about that issue.

Arnold also asks how market failure affects my claim that business can be a way of exercising beneficence. He agrees that business practice can be beneficent in competitive markets free of externalities and other market failures. But he wonders what we should say about real-life markets, which are often not fully competitive and suffer from market failures.

Note that the concept of "market failure" is rigoristic. A nearly perfect market—a market that is 99.999999% Pareto-efficient—nevertheless is said to exhibit market failure, as the concept is defined in economic theory. But neither Arnold nor I think that small market failures imply that businesses cannot be beneficent. Remember, all the stories I told about businesses alleviating poverty occurred in "failed"—that is, *real*—markets.

My aim here is not to give business a free pass. It's instead to stop holding business to much higher standards than we hold government, charities, NGOs, volunteering, and the like. Consider, for instance, that all NGOs, governments, charities, and volunteering organizations also produce externalities—for instance, pollution. But we don't say that Against Malaria automatically fails to be beneficent because it contributes to climate change. We don't say that government action is never beneficent even though no governments are 100% Pareto-efficient. We don't say that nurses cannot be beneficent even though nursing has principal-agent problems, asymmetric information, and the like.

We should treat government, NGOs, charities, civil society, social business, and regular business by the same basic standards.

In the real world, organizations and agents in these sectors have a mix of motives. Some do better work than others. It's an open question whether any particular soldier, president, charitable worker, volunteer, ESG investor, venture capitalist, or nurse makes the world a better place, or whether their particular actions count as beneficent or not. If we study many such people, we might be able to make some generalizations, such as that the typical medical doctor saves about two lives per year, or that the typical Russian government bureaucrat does more harm than good. I suspect most businesses in the West make the world a better place, though they all create negative externalities (as all real entities do). I am not saying business is special; I am saying it isn't. Once we realize that business isn't special, we realize that business is a good way to discharge many of our duties of beneficence.

Response to Chappell

I suspect most people should give more to charity than they do, even though I measure beneficence not by amount of money donated, but by amount of good done. I am unsure what the cut-off is for any particular person to have "done enough," but I suspect many of us fall below that line. More importantly, I suspect most people should give *better* rather than more.

In my original remarks, despite my enthusiasm for EA charities, I claimed that charity has no record of creating sustained economic development of the kind that turned, say, South Korea from poor to rich. I can *imagine* scaled-up private charities producing such development, but I doubt in practice they would, even with trillion-dollar budgets. Why not?

The economist Angus Deaton notes that eliminating poverty *seems* like it's a simple problem of distribution:

> One of the stunning facts about global poverty is how little it would take to fix it, at least if we could magically transfer money into the bank accounts of the world's poor. In 2008, there were about 800 million people in the world living on less than $1.00 a day. On average, each of these people is "short" about $0.28 a day; their average daily expenditure is $0.72

instead of the $1.00 it would take to lift them out of [extreme] poverty. We could make up that shortfall with less than a quarter billion dollars a day ... Taking ... into account [differences in purchasing power in poor countries] ... world poverty could be eliminated if every American adult donated $0.30 a day; or, if we could build a coalition of the willing from all the adults of Britain, France, Germany, and Japan, each would need to give only $0.15 a day.[6]

Deaton adds:

These calculations ... are examples of what I call the *aid illusion*, the erroneous belief that global poverty could be eliminated if only rich people or rich countries were to give more money to poor people or to poor counties ... far from being a prescription for eliminating poverty, the aid illusion is actually an obstacle to improving the lives of the poor.[7]

Here, Deaton means to critique government-to-government foreign aid. Like many economists, he thinks the track record of such government aid is poor—it mostly retards development, in part by supporting corrupt, extractive governments and preventing the institutional reform needed to spur real growth.

But I suspect Deaton's remarks also apply to private charity. Eliminating poverty is not just about giving poor people enough money to get them over some line. If the only reason Bob eats is that I feed him, Bob would starve if I stopped. The real goal is to get him to a position where he doesn't need to be fed by others.

Thus, the more important goal is to create such a high levels of development that normal workers are rich without charity. For instance, thanks to past capital accumulation, good management techniques, and inclusive government institutions, it's easy for an American with a normal job to live well. In contrast, thanks to badly functioning, extractive institutions, and low capital development, a person living in Haiti who works as hard as he can and exerts extraordinary produce will nevertheless expect to remain poor.

Ending poverty requires development. We know that business investment under stable, inclusive government institutions creates that development. Charity can be a part of that solution. Charity stops people from dying and reduces suffering while we wait for development to occur. But we don't have record of it yet substituting for or causing development.

I doubt that altruistically-motivated people with a trillion-dollar budget would know how to create such development. The problem here isn't a budget problem, the information problem that all central planners face.[8] Consider that command-economy socialism universally failed because, in the absence of market forces, planners could not determine what to produce, where, in what quantities, at what quality, and so on. Planners had to plan everything all at once. That task can't be done, so it wasn't. In contrast, markets spread planning out to all market participants, using prices (which emerge from supply and demand) to communicate information, all without most market planners even knowing what prices mean. Asking hypothetical NGOs with trillion-dollar budgets to create development is too much like asking socialist planners to create sustained development. It *could* work, but it almost certainly won't.[9] So, when it comes to fighting poverty. Charity has a role, but substituting for markets or creating long-term development is not that role.

Response to Davis

Davis wonders whether we have any general duties of beneficence. He instead thinks that beneficence is above and beyond the call of duty.

Davis asks us to imagine a person—let's call him Herman the Hermit—who chooses to live alone in the woods, cut off from other humans. To flesh out the example, imagine that Herman did not first discharge all his positive duties to others (if he has any) before retiring to live alone. Instead, imagine that right as Herman graduates from high school, he waves goodbye to his parents, buys some equipment, and leaves to live alone and without human interaction the rest of his days. Let's imagine that while Herman prefers to be a hermit, it's a weak preference. He's not choosing to be a hermit because he can't stand other people and finds social interaction horrible.

Herman does not repay the taxes that paid for his education. Nor does he take care of his parents in his old age. He does not volunteer for Habitat for Humanity or raise money for Give Directly. He never saves strangers left stranded on the side of the road. He does nothing to help the poor or alleviate the suffering of those in need.

I agree with Davis: it does not seem that Herman is violating any moral duties, even though he never helps others, never repays the taxes that were used to educate him, and so on. Herman is no hero or moral exemplar, but it doesn't seem like he's doing anything morally wrong. I don't want to claim that morality conscripts Herman into being a social animal, forcing him to live with others in order to serve them.

We can present Davis's argument in numbered premise form:

1. It's permissible to become an isolated hermit.
2. But an isolated hermit will never act beneficently.
3. Therefore, it's permissible never to act beneficently.

Premise 1 shouldn't be interpreted as saying anyone may choose to become an isolated hermit at any time. For instance, I no longer may—I have special obligations to my spouse and children, an enforceable mortgage, and contractual obligations to Georgetown to work at least one more year before I may quit. But Davis is having us imagine a person with no special, specific duties like that.

One might try to get around this argument by rejecting the inference. Maybe the conclusion follows only if read it in a very specific way: Perhaps hermits have no duty to act beneficently because they are outside society, but perhaps people who choose to remain in society do.

But that move seems weird. Beneficence is about helping people, including strangers, because of their need. In contrast, reciprocity is about repaying others for their good deeds, civic virtue is about promoting the common good of one's society, fairness is partly about avoiding free riding, and so on. So, there's nothing about beneficence per se that suggests it arises as a duty from social interactions or relationships with others. Indeed, as an Effective Altruist, I think the most choice-worthy targets of my beneficence are the world's worst-off people, though they are members of my community and are not much involved in my extended networks of trade or mutual advantage.

Thus, I admit Davis has a very strong counterexample to the commonsense view that we possess general obligations of beneficence. Perhaps, as he claims, pure beneficence can be a good thing, but it is not a duty. This allows that we can have special obligations to help specific people, as in the special obligations I owe my children, spouse, or students. Our relationships might give rise to obligations of reciprocity and community. So, perhaps Davis does not mean to suggest that we never have positive obligations in the vicinity of beneficence. Still, if it can be permissible to be a hermit, then there is at least one case where people who could help others can simply choose not to do so, without blame or wrongdoing. How much does that generalize? If Herman doesn't have to help others (out of beneficence), why must I?

Notes

1 A nihilist says we have no obligations, so it's trivial to properly weight all legitimate interests, properly respect all rights, etc. An ethical egoist says that your only basic obligations are to yourself.
2 See Hasnas 2013 for a review of how stakeholder theory is, by the lights of the very people who originated it, so vacuous that even purportedly rival theories instead count as instances of it.
3 For example, consider Arnold and Bowie 2003, a famous paper cited hundreds of times and placed in dozens of anthologies. This paper starts with the Kantian slogan that we should treat all people as ends in themselves and not mere means. It then moves to the more substantive view that we should promote their autonomy. Fair enough. But it then concludes that this means they must be paid a living wage. Where did that conclusion come from? No actual argument is given.
4 Bazerman and Tenbrunsel 2011.
5 For a critique, see Alcott et al. 2023. This paper argues that businesses do a lot more good by generating consumer surplus than they do by trying to promote ESG and other ideas.
6 Deaton 2013, 268–269.
7 Ibid., 270.
8 Hayek 1945.
9 See https://blog.givewell.org/2009/10/26/helping-farmers-is-harder-than-youve-heard/ for a critique of Heifer International along these lines.

PART III
Exploring Beneficence

Richard Yetter Chappell

11
EXPLORING BENEFICENCE

Richard Yetter Chappell

Three Questions Exploring Beneficence

Something I find exciting about Effective Altruism is the way that it exemplifies a deliberate, rational attempt to do more good *as such*. This is surprisingly rare. Many people are drawn to particular good causes. But strikingly few have pursued the project of *moral prioritization*, weighing the tradeoffs between all the various worthy causes out there, to try to determine which should be our *top* priority. Such judgments are always fallible, of course, and open to revision in light of new evidence and arguments. But rationally pursuing a difficult goal seems likely to get us closer to achieving it than would not even trying, so I consider it a good and worthy project.

An implicit assumption of Effective Altruism is that reasons of *beneficence*—or doing good—are among our strongest moral reasons. And Effective Altruists tend to have a distinctive, "utilitarian-esque" understanding of what our reasons of beneficence involve: roughly, *improving global well-being*. As someone with strong utilitarian sympathies myself, I've crafted my three questions with an eye to clarifying and exploring this distinctive conception of beneficence.

My first question asks whether EAs are right to ascribe such importance to our reasons of beneficence. I argue that they are, and more radically that we can assess how morally good a person is (to a first approximation) simply by looking at how much they (want to) do to improve global well-being.

My second question, on beneficence and self-sacrifice, looks at how acting to *promote the good* can diverge from ordinary conceptions of "altruistic" behavior (which place more weight on self-sacrifice than on actually helping others). In my essay, I argue that the ordinary conception illicitly conflates *being* good with *looking* good. So I think the EA conception is a clear improvement.

Finally, my third question asks how much we should care about future generations. The "longtermist" branch of EA has courted controversy by suggesting that slight reductions to the risk of human extinction can morally outweigh saving lives for certain. This is a departure from ordinary assumptions about how we should help others. But again, I argue that these revisionary verdicts are well warranted in principle.

Question 7: How Important Is Beneficence Compared to Other Values?

Beneficence consists in helping others, such as by saving and improving lives. How important or worthwhile is this, compared to other values we might either honor or promote through our actions?[1] My answer is that beneficence is, generally speaking, the *most* important value for guiding our moral lives. Other values are important when they help promote social coordination and thereby serve overall well-being.[2] But if we hold all else equal, their *non-instrumental* significance pales in comparison to the significance of saving and improving lives.

What Makes for a Better Cause?

People generally like the sound of beneficence. (Who could be *opposed* to saving and improving lives?) Giving to charity is sometimes thought to be the paradigm of ethical action. But it doesn't rile up our emotions the way that *fighting against injustice* does. As political animals, we like to define ourselves in contrast to a perceived out-group. And denouncing our political enemies and their associated vices (racism! sexism! transphobia!) is ever so much more satisfying than simply saving lives. So, while we like the sound of beneficence, it's rare to give it more than lip service.

Too often, our hearts belong to the culture war. Or environmentalism. Or some other form of local social justice activism.[3]

I think this is a moral mistake. However important our local problems may be, they cannot compare to the importance of saving lives on a global scale. People are *literally dying* for want of a few thousand dollars invested in anti-malarial bed nets and other cheap and simple interventions.[4] There's something very wrong with the fact that this takes up so little of our collective moral concern and attention.

Of course, one could always justify an alternative focus by arguing that one's efforts there do even *more* good in expectation. I don't mean to rule out this possibility: if you've truly found a cause that better serves overall well-being, that's great! But we should be open to the possibility—indeed, the *likelihood*—that the most emotionally appealing causes are not actually the most morally important. To test our gut preferences, we can assess the relative moral importance of different permissible options by way of their expected impact on global well-being, impartially considered. We have strong moral reasons to want to save and improve more lives, and hence to focus our moral efforts on the (permissible) causes where we would have the greatest positive expected impact.[5]

What Makes for a Better Person?

A good moral agent will prioritize morally more important ends over less important ones. This suggests that we can assess how morally good a person is (to a first approximation) simply by looking at how much they (want to) do to improve global well-being.

We may split this assessment into two dimensions. How *virtuous* or well-meaning a person is depends on their desires, the expression of which depends on contingencies of their life situation. A billionaire who donates one million dollars to charity is much less generous, in disposition, than a poorer person who donates a large portion of her income and would give *many* millions were she as wealthy as the billionaire. But how morally *beneficial* a person is to the world simply depends on the net value of their contributions: in giving a million dollars, the stingy billionaire presumably does more good than most of the rest of us are able to.

Both dimensions of agential assessment can be approximately reduced to assessments of beneficence, just in different ways. One measures the strength of a person's beneficent desires; the other, how much they actually contribute to promoting beneficent ends.

Either suggestion is apt to seem somewhat revisionary. We typically assess people's moral characters in more interpersonal terms. Roughly: *do they seem like a loyal and reliable ally?* We like and approve of people who are good friends, spouses, parents, and colleagues. We may be more impressed by someone who comes across as warm, kind, and helpful in personal interactions, than by one who determinedly works extra hours to be able to donate more to charity.

This makes sense if our assessment is self-interested at heart: the former person may well make a better friend for you! But it's puzzling if the assessment is meant to be a *moral* one. Why think that being lovely to be around is morally more important than saving others' lives? (Maybe life wouldn't be worth much if it weren't for interpersonal warmth. But there's no general lack of that. We can trust that the people whose lives would be saved would find themselves surrounded by loving friends and family. The value of interpersonal warmth may, like the value of health, be among those that are well promoted by general beneficence.)

It may be that a common explanation for our failure to recognize the central importance of impartial beneficence is simply that we, ourselves, are not sufficiently impartial or beneficent. We would sooner assess people in terms of how good they are likely to be *to us*. Such a focus may be instrumentally rational for many purposes. But we can see on reflection that *moral* assessment is different, and calls for a more impartial standard.

I should flag that my suggested exclusive focus on beneficence is just an approximation. We can imagine an evil billionaire who donates by day and tortures the oppressed by night. Even if the good done slightly outweighs the bad in aggregate, so they are (by stipulation) beneficial to the world on net, we needn't consider such a monster *virtuous*. Delighting in harm to others reveals an unusually malicious character, after all.[6] But such cases are surely rare. Most people presumably meet the minimal standards for *not being evil*; what's left unsettled is just *how good* we're going to be. Most will happily settle for being OK, and that's OK (not great, but not terrible either). What I'm suggesting here is just that many of the things we typically associate with moral motivation or good character aren't really going to make that much of a difference from this point. By far the greatest factor—beyond not being evil—is just *how beneficent* one is.

So if you want to be a better person, the place to focus is on trying to do more good, impartially considered. And that just is the project of effective altruism: trying to do good, effectively.

Question 8: Does Self-Sacrifice Make Beneficence More Virtuous?

Charity and volunteering involve an element of self-sacrifice. Does that make these acts more virtuous than improving the world via other means (such as one's career)? Do the possible reputational rewards of public, ostentatious philanthropy provide us with moral reasons to donate anonymously instead? In this essay, I distinguish *moral reasons for action* from *evidence of virtue*, and suggest that we should not be overly concerned about the latter. It is even most virtuous to simply want to do the most good, rather than being overly concerned about whether one's actions reliably *signal* altruism or virtue. So even virtue ethicists have reason to encourage a more outward-looking practical outlook.

Reasons and Virtue-Signals

Distinguish (1) doing the right *thing*, (2) acting *for the right reason*, and (3) acting in a way that *others can see* is for the right reason. The first is acting ethically. The second is acting virtuously, esteemably, or with moral worth. And the third *signals* virtue.[7]

I suspect that some common views about acting ethically are better understood as views about how to reliably signal virtue. When virtue-signaling conflicts with actually doing good, such signaling becomes morally vicious. Given a choice between *actually* helping others or *looking like* we're helping others, a truly virtuous person will obviously prefer the former. Indeed, an ideally virtuous person would not care about their personal reputation at all (except insofar as it may indirectly help them to more effectively help others).

The distinction between the morality of actions and intentions is a familiar one. We can all imagine someone doing the right thing for the wrong reasons (e.g., saving a drowning child in hopes of getting featured in the news), or performing a morally harmful action from the best of intentions (e.g., saving a drowning child who turns out to be the next Hitler). Often, good acts could plausibly be done from a variety of motives, some more morally laudatory than others. Cynical observers can often hypothesize self-interested reasons that might underlie the actions of an apparent do-gooder. As a result, we cannot simply infer virtue (or good intentions) from ethical action. But because observers are often especially interested in the quality of agents' moral motivations, they may turn to heuristics that more reliably signal virtue (such as costly signals of altruism). Because virtue-signaling is not the same as actually acting well, this move is morally distorting. By focusing overly much on costly signals of virtue, we lose sight of what really matters—namely, helping others.

In what follows, I'll step through two examples of this moral distortion: the valorization of anonymous donation, and the valorization of self-sacrifice. I argue that we often have strong moral reasons to prefer both public giving and mutual benefit. Without self-sacrifice, observers cannot so easily tell whether the agent's motivations are truly altruistic. But that is less important than actually helping more. And since the *truly* virtuous agent is moved by what's really important, it turns out that suboptimal philanthropic "virtue signaling" is incompatible with genuine virtue.

Virtue and Anonymous Donation

It's commonly thought to be more virtuous to give to charity anonymously rather than publicly trumpeting one's donations. (That way, the thought goes, you ensure that your motivations are pure.) And it is true that there would be something morally distasteful about a donor who cared more about boosting their own reputation than about actually helping others. Such a person would not be acting for the right reasons. But nor would a donor who remains anonymous for fear of seeming a braggart. The truly virtuous agent simply cares about helping others, and will take whatever permissible means best serve this end.

As it happens, others are better helped by our publicizing our philanthropy, as this promotes social norms that prompt others to give more too. As Peter Singer explains:[8]

> One of the most significant factors determining whether people give to charity is what others are doing. Those who make it known that they give to charity increase the likelihood that others will do the same ... We need to get over our reluctance to speak openly about the good we do. Silent giving will not change a culture that deems it sensible to spend all your money on yourself and your family, rather than to help those in greater need.

There may in fact be personal costs to publicizing one's philanthropy, due to the phenomenon of 'do-gooder derogation'—whereby the morally mediocre conspire to ward off threats to their moral self-conception.[9] An agent who persists despite these personal costs, because they care even more about helping others via promoting a "culture of giving," is thereby acting *especially* selflessly (though others may not be in a position to know this). Another donor who allows themselves to be bullied into silence about the value of philanthropy would not only do less good, they would also be less virtuous. Someone who learned of the second agent's anonymous donation might mistakenly believe them to be especially virtuous for having donated anonymously, but they would be mistaken in this. It would have done more good to publicize their giving. And it is more virtuous to be motivated to do more good.

As this example shows, trying to do more good (and hence *actually* being virtuous) may prompt us to act in ways that systematically diverge from conventional assumptions about virtue. The latter seem to more closely match how one would act if trying to *signal* virtue (at least to oneself); except that, as we have now seen, it does not even reliably do that. To *truly* signal virtue, you would need to signal that you are simply motivated to *do what most helps others, regardless of what signals this sends*. But if the acts that most help others could often be performed from a variety of motives, there may not be any straightforward way to signal that one did the action from the correct motives. Of course, a truly virtuous agent won't care about that. They just want to help others. Alas, we real-life humans tend to have more complex motivations. But whatever our actual mess of motivations may be, it could be worth trying to de-emphasize them, and focus on the moral world beyond our own minds. That's what the virtuous agent would do, after all.

Altruism and Beneficent Careers

Could volunteering at the local soup kitchen be a bad idea? Well, consider the opportunity costs. A medical researcher working on a new vaccine that could save millions of lives is likely to do vastly more good through their regular work. Or consider a lawyer or entrepreneur who could earn more through working longer hours, and then donate their extra wealth to likewise do vastly more good. For highly skilled professionals, it would be rather incredible if doing *unskilled labor* was really the best use of their time. Yet volunteering is widely valorized as a way to *show that we care*. I would prefer for people to think less about how to signal caring, and more about how they can *best use their skills* to actually make the world a better place.[10]

I don't mean to claim that volunteering is never worthwhile. But it makes sense to prioritize work that uses your distinctive skills over labor that doesn't. Our well-meaning lawyer could surely do a lot more good working *pro bono* for deserving clients than ladling soup that could just as well be ladled by a high-school dropout.

This contrast is apt to strike many as offensive. Perhaps part of the appeal of volunteering is its egalitarian ethos: no matter how in-demand your skills may be in the professional world, everyone in the soup kitchen is more or less equal when it comes to the task at hand. And perhaps there are moral benefits to slowing down and providing basic services to those in need, face-to-face. But there are also moral benefits to saving more lives, so we can't just assume that the conventional recommendation wins out as "overall" for the best.

We should be open to the possibility that working more and volunteering less could be the morally better option. The fact that a scientist gets paid for their life-saving research doesn't make it inherently less moral or worthwhile. (Ideally, price signals serve to indicate where our labor could most productively be used. Although obviously in practice this doesn't correlate perfectly with beneficent impact, the correlation isn't *negative*, either.) Again, we may be misled by the heuristic of using *selflessness* as a signal of virtue. But this is distorting. We should simply want to do good, not to do good *and suffer for it*. If we can do good through skilled work that others are willing to pay for, all the better.

Question 9: How Much Should We Care About Future Generations?

How much should we care about future people? Total utilitarians answer, "Equally to our concern for presently-existing people." Narrow person-affecting theorists answer, "Not at all"—at least in a disturbingly wide range of cases.[11] I think the most plausible answer is something in-between.

Person-Directed and Impersonal Reasons

Total utilitarianism is the view that we should promote the sum total of well-being in the universe. In principle, this sum could be increased by either improving people's lives or by adding more positive lives into the mix (without making others worse off). I agree that both of these options are good, but it seems misguided to regard them as *equally* good. If you see a child drowning, resolving to have an extra child yourself is not (contra total utilitarianism) an adequate substitute for saving the existing child. In general, we're apt to think, we have stronger reasons to make people happy than to make happy people.[12]

On the other hand, the narrow person-affecting view can seem disturbing and implausibly extreme in its own way. Since it regards happy future lives as a matter of moral indifference, it implies that—if it would make us happier—it'd be worth preventing a future utopia by sterilizing everyone alive today and burning through all the planet's resources before the last of us dies off. Utopia is no better than a barren rock, on this view, so if faced with a choice between the two, we've no moral reason to sacrifice our own interests to bring about the former.

Our own value—and that of our children—are seen as merely *conditional*: given that we exist, it's better to make us better-off, just like *if* you make a promise, *then* you had better keep it.[13] But there's no reason to make promises just in order to keep them: kept promises are not *in themselves* or unconditionally good. And narrow person-affecting theorists think the same of individual persons. Bluntly put: we are *no better than nothing at all*, on this bleak view.

Fortunately, we do not have to choose between total utilitarianism and the narrow person-affecting view. We can instead combine the life-affirming aspects of total utilitarianism with extra weight for those who exist antecedently.[14] On a commonsense hybrid approach, we have both (1) strong person-directed reasons to care especially about the well-being of antecedently existing individuals, and (2) weaker impersonal reasons to improve the world by bringing additional good lives into existence. When the amount of value at stake is sufficiently large, even reasons of the intrinsically weaker kind may add up to be very significant indeed. This can explain why avoiding human extinction should be a very high priority on a wide range of reasonable, life-affirming views, without depending on anything as extreme as total utilitarianism.

In Defense of Good Lives

There are three other common reasons why people are tempted to deny value to future lives, and they're all terrible. First, some worry that we could otherwise be saddled with implausible procreative obligations. Second, some think that it allows them to avoid the paradoxes of population ethics. And, third, some are metaphysically confused about how non-existent beings could generate reasons. Let's address these concerns in turn.

Imagine thinking that the only way to reject forced organ donation was to deny value to the lives of individuals suffering from organ failure. That would be daft. Commonsense morality grants us strong rights to bodily integrity and autonomy. However useful my second kidney may be to others, it is *my body*, and it would be supererogatory—above and beyond the call of duty—for me to give up any part of it for the greater good of others.

Now, what holds of kidneys surely holds with even greater stringency of uteruses, as being coerced into an unwanted pregnancy would seem an even graver violation of one's bodily integrity than having a kidney forcibly removed.[15] So recognizing the value of future people does not saddle us with procreative obligations, any more than recognizing the value of dialysis patients saddles us with obligations to donate our internal organs. Placing our organs in service to the greater good is above and beyond the call of duty. This basic commitment to bodily autonomy can survive whatever particular judgments we might make about which lives contribute to the overall good. It does not give us any reason to deny value to others' lives, including future lives.[16]

The second bad argument begins by noting the paradoxes of population ethics, such as Parfit's "Mere Addition Paradox," which threatens to force us into the "Repugnant Conclusion" that any finite utopian population A can be surpassed in value by a sufficiently larger population Z of lives that are barely worth living.[17] Without getting into the details, the mere addition paradox can be blocked by denying that good lives are absolutely good at all, and instead regarding different-sized populations as *incomparable* in value.

But this move ultimately avails us little, for two reasons: (1) it cannot secure the intuitively desirable result that the utopian world A is *better* than the repugnant world Z; and (2) all the same puzzles about quantity-quality tradeoffs can re-emerge *within* a single life, where it is not remotely plausible to deny that "mere additions" of future time *can* be of value or increase the welfare value of one's life. Since we're all committed to addressing quantity-quality tradeoffs within a life, we might as well extend whatever solution we ultimately settle upon to the population level too. So there's really no philosophical gain to temporarily dodging the issue by denying the value of future lives.

The third argument rests on a simple confusion between absolute and comparative disvalue. Consider Torres:[18]

> [T]here can't be anything bad about Being Extinct because there wouldn't be anyone around to experience this badness. And if there isn't anyone around to suffer the loss of future happiness and progress, then Being Extinct doesn't actually harm anyone.

I call this the 'Epicurean fallacy,' as it mirrors the notorious reasoning that death cannot harm you because once you're dead there's no longer anyone there to be harmed. Of course, death is not an absolutely *bad state* to be in (it's not a state that you are ever in at all, since to be in a state you must exist at that time). Death's intrinsic neutrality instead makes you worse off *in comparison* to the alternative of continued positive existence. And so it goes at a population level: humanity's extinction, while absolutely neutral, would be awful *compared* to the alternative of a flourishing future containing immensely positive lives (and thus value). If you appreciate that death can be bad—even tragic—then you should have no difficulty appreciating the metaphysical possibility that extinction could be even more so. (Though we can imagine worse things than extinction, just as we can imagine worse fates than death.)

An Agnostic Case for Longtermism in Practice

William MacAskill defines longtermism as "the idea that positively influencing the longterm future is a key moral priority of our time."[19] After all, the future is vast. If all goes well, it could contain an astronomical number of wonderful lives. If it goes poorly, it might soon contain no lives at all—or worse, overwhelmingly miserable, oppressed lives. Because the stakes are so high, we have extremely strong moral reasons to prefer better long-term outcomes.

That in-principle verdict strikes me as difficult to deny. The practical question of what to do about it is much less clear, because it may not be obvious what we can do to improve long-term outcomes. But longtermists suggest that there is *at least* one clear-cut option available, namely: *research the matter further*. Longtermist investigation is relatively cheap, and the potential upside is immense. So it seems clearly worthwhile to look more into the matter.

MacAskill himself suggests two broad avenues for securing positive longterm impact: (1) contributing to economic, scientific, and (especially) moral progress—such as by *building a morally exploratory world* that can continue to improve over time; and (2) working to mitigate existential risks—such as from nuclear war, super-pandemics, or misaligned artificial intelligence—to ensure that we have a future at all.

This all seems very sensible to me. I personally doubt that misaligned AI will take over the world—that sure doesn't seem the most likely outcome. But a bad outcome doesn't have to be the "most likely" in order for it to be prudent to guard against. I don't think any given nuclear reactor is likely to suffer a catastrophic failure, either, but I still think society should invest (*some*) in nuclear safety engineering, just to be safe.[20] Currently, the amount that our society invests in reducing global catastrophic risks is negligible (as a proportion of global GDP). I could imagine overdoing it—e.g., in a hypothetical neurotic society that invested the majority of its resources into such precautionary measures—but, in reality, we're surely erring in the direction of under-investment.

So, while I don't know precisely what the optimal balance would be between "longtermist" and "neartermist" moral ends, it's worth noting that we don't need to answer that difficult question in order to at least have a *directional* sense of where we should go from here. We should not entirely disregard the long-term future: it truly is immensely important. But we (especially non-EAs) currently *do* almost entirely disregard the long-term future. So it would seem wise to remedy this.

Notes

1 We promote a value by bringing about more of it, in consequentialist fashion. We respect a value by constraining our actions so as to avoid violating the value ourselves, in non-consequentialist fashion. The two approaches may conflict in theory, for example, if we're in a position to prevent five killings by means of killing one innocent person ourselves. Killing one to save five serves to *promote* the value of human life, but does not *respect* it (in the above sense). See Pettit 1991.
2 We should strive to be honest and trustworthy, for example, precisely because this helps to preserve the social conditions necessary for promoting overall well-being over the long run.
3 Perhaps an even larger number of people are morally apathetic, and aren't especially trying to do any kind of good at all. But the suboptimality of moral apathy seems relatively trivial, so here I focus instead on those with strong—but possibly misdirected—apparent moral motivations.
4 See www.givewell.org.

5 I limit this claim to just *permissible* options and causes so as to remain compatible with non-consequentialist constraints on the pursuit of good outcomes. That makes no practical difference here—no EA organization that I'm aware of proposes that we kill people to harvest their organs, or anything like that. On the contrary, EA causes are clearly permissible. A more practically significant point of contention instead concerns whether a permissible cause's being *impartially better* means that it is also *morally* better than more parochial (but emotionally satisfying) forms of moral expression. Effective Altruism urges us to prioritize the impartially best of our permissible options, whereas its critics push back against this demand.

6 We may even think poorly of one who callously neglects those drowning on their doorstep while helping distant others, though whether such disapproval is warranted may depend on further details of the agent's psychology. See Chappell and Yetter-Chappell 2016, and Chappell 2019b.

7 Which is not to say that the agent has necessarily done it *for* that reason. An act may signal—or provide evidence of—virtue without the agent intending for this to be so.

8 "Look who's being generous!" *The Guardian*, June 22, 2008, Available at: http://www.guardian.co.uk/money/2008/jun/22/charitablegiving (accessed July 27, 2023).

9 Minson and Monin 2012. See also Schwitzgebel 2019.

10 See 80000hours.org for career advice from an Effective Altruist perspective.

11 They allow that we shouldn't want future individuals to suffer. And they allow that we should prefer any given future individual to be better off rather than existing in a worse-off state. But they think we have no non-instrumental reason to want the happy future individual to exist at all. And also no non-instrumental reason to prefer for a happier individual to exist *in place of* a less well-off, alternative future person. For a general introduction to population ethics, see "Population Ethics," in Chappell, Meissner, and MacAskill 2023.

12 Narveson 1973, 80.

13 Frick 2020.

14 McMahan 2013; Chappell 2017.

15 For related analogies, see Thomson 1971.

16 This basic argument is further developed in Chappell 2017.

17 Parfit 1984.

18 Torres 2023a, 203.

19 MacAskill 2022, 4.

20 Of course, that's not to endorse pathological regulation that results in effectively promoting coal power over nuclear, or other perverse incentives.

12
ARNOLD'S RESPONSE TO CHAPPELL

Chappell argues for three claims.

First, beneficence, which "consists in helping others, such as by saving and improving lives," is "the *most* important value." Relatedly, "we can assess how morally good a person is (to a first approximation) simply by looking at how much they (want to) do to improve global well-being."

Second, the morally important thing is to *do* good, not *look* good. We should prioritize doing the right thing over *signaling* that we're doing the right thing.

Third, future people matter, morally speaking, and safeguarding a positive future for humanity should be "a key moral priority of our time." To this end, we should continue to promote economic, scientific, and moral progress, as well as work to mitigate existential risks.

Before turning to the first and third claims, let me quickly register agreement with Chappell's second claim. Too many people, Chappell argues, care about *seeming* good, when really they should focus on actually *doing* good. Strangely, actually doing good "may prompt us to act in ways that systematically diverge from conventional assumptions about virtue." For instance, most people laud the anonymous donor over the named one; I know I certainly did before reading Chappell! And yet, if "others are better helped by our publicizing our philanthropy, as this promotes social norms that prompt others to give, too," then the truly good donor will advertise his beneficence far and wide—even if this means running the risk of *seeming* a braggart. Or again: the

DOI: 10.4324/9781003508069-16

truly good agent will allocate her labor inputs with an eye toward promoting global well-being, *even if* this means eschewing socially-lauded, but low-impact volunteer work. A scientist may do more good grinding in the lab than ladling soup in a shelter.

Again, I find all of this very convincing; Chappell's second entry has me persuaded. I'm less persuaded by his first entry. And about his third entry, I don't know *what* to think—but I have some questions.

Beneficence as the Master Virtue?

Some moral theories provide an account of right *action*. Others provide an account of good *character*. The first sort of theory tells us how to *act*, whereas the second tells us how to *be*. Chappell's theory collapses these categories, defining good character in terms of right action. That is: we should try to *be* the sort of person who acts *rightly*, i.e., who promotes good outcomes, impartially considered.

Chappell further divides agential assessment into two dimensions. How *virtuous* a person is depends on how much she wants to promote the good. So, it's a matter of her psychology. How *morally beneficial* a person is depends on how much good she actually produces. These dimensions can come apart, as when a well-meaning person chooses inefficient means to her virtuous end of promoting global well-being; or a malevolent person stumbles into beneficent action by accident.

In sum, Chappell argues that if you want to be a morally good person—and who doesn't?—you should *increase your desire to do good*, and then *act* on that desire in a sensible, efficient way. In other words, you should be an Effective Altruist, since "trying to do good, effectively" just is the EA project.

Chappell concedes that this theory of agential assessment is "apt to seem somewhat revisionary." That, I think, is putting it mildly! In truth, Chappell's theory departs *radically* from standard modes of character assessment.

Consider the best people you know: the sorts of people you admire the most, the people you'd hold up as good moral exemplars for your kids or even for yourself. What distinguishes them? I bet they possess some combination of the following traits: kindness; generosity; a sense of humor; a sunny disposition; perseverance and resilience; trustworthiness; courage and integrity; wisdom.

Notice that most of these traits have nothing to do with promoting the impartial good. That is, they have nothing to do with the trait, benevolence, that Chappell says is *the* defining feature of the good person. A sense of humor, wisdom, courage, integrity, and more: these are things we'd all want for our children, these are things we all admire in other people, and yet they go entirely missing from Chappell's account of good character.

And even kindness and generosity, which are the closest matches to Chappell's master virtue of benevolence, don't really pick out precisely the same idea as benevolence. Importantly, most people intuitively like and admire generous and kind people, but I'm not sure they feel so warmly about benevolent people as such.

After all, benevolence is *impartial*; it instructs us to promote overall well-being, without regard to whose well-being it is. So, a benevolent father will not necessarily dote on his own children, given the opportunity costs for overall well-being of said doting. (Time tossing the pigskin with junior could be more efficiently spent earning money to donate to EA causes, etc.) A *kind* father would dote, but a *benevolent* one might not. Or again, a *generous* father or friend will direct at least some of that generosity toward those near and dear; but a *benevolent* father or friend probably wouldn't, since $100 for junior's birthday is a bad play, utility-wise, relative to other donative options.

In sum, Chappell's moral exemplar is not most people's moral exemplar. That's hardly surprising, I suppose, given Chappell's background consequentialism. That consequentialists make poor friends, husbands, and wives, even if good cosmopolitans—good citizens of the world—is, admittedly, an old critique. I don't purport to be claiming anything new here! But still, I think the critique applies to Chappell's account. To put a fine point on my objection: isn't Chappell's moral exemplar, the beneficent agent, actually a deeply alien and unappealing character, someone whom basically no one outside a philosophy seminar room would ever want to be, or befriend, or marry?

To be clear, I am open to the idea that beneficence is *one* virtue among many in the moral exemplar's repertoire. But that, I take it, is not Chappell's claim. Instead, he holds that "if you want to be a better person, the place to focus is on trying to do more good, impartially considered," as opposed to, say, being more kind, or considerate, or truthful, or brave, or funny, or whatever. For Chappell, benevolence really is the master—and perhaps, in effect, the only—moral virtue. And again, that seems implausible to me.

One last point. Chappell considers "an evil billionaire who donates by day and tortures by night." Must we call him virtuous? No, Chappell says, because even though he (let's stipulate) produces more good than bad, "delighting in harm reveals an unusually malicious character." I think Chappell lets himself off the hook a little too easily here. Let's make the case harder. The billionaire doesn't *delight* in torturing; instead, he tortures to rejuvenate, and nothing else would serve. Thus refreshed, and *only* thus refreshed, he wakes each morning ready to amass astounding piles of money, which he dutifully donates to EA-approved causes. The picture, then, is of a man utterly devoted to the impartial good; he just has this tragic quirk, such that to maximize the good overall, he has to infringe it locally, in the form of a little light torture.

I say this man is a monster—a utility-maximizing one, but a monster nonetheless. I see no resources in Chappell's theory for reaching that same verdict.

Longtermism

"Longtermism," Chappell explains (citing MacAskill), is "the idea that positively influencing the longterm future is a key moral priority of our time." Chappell endorses longtermism. I *think* I do, too, and yet I remain troubled by some of its implications. In what follows, I'll sketch the consequentialist case for longtermism, then present (one aspect of) Torres's powerful critique of longtermism, and conclude by asking Chappell how he'd respond to Torres's critique.

I begin with the consequentialist argument for longtermism. From the consequentialist perspective, the future is of profound moral importance. The reason is simple. Consequentialism says to maximize well-being, including human well-being. But "almost all of humanity's life lies in the future."[1] It follows that almost all potential human well-being lies in the future as well.

Humanity has been around for the merest instant, relative to the Earth's habitable period. Provided we don't go extinct, humanity can expect to inhabit the Earth for another billion years. After that, the sun, having exhausted its nuclear fuel, will turn into a red giant, swallowing Mercury, Venus, and, sadly, the Earth.

But consider two reasons for optimism. First, perhaps we'll colonize the stars, thereby decoupling our survival from Earth's. Second, even if our species goes down with the ship, we'll have had a very good run. Astronomer Carl Sagan estimates that some 500 *trillion* humans will have lived prior to the sun's demise. EA philosopher Nick Bostrom puts the number 20 times higher: 10 to the 16th power, or 10,000 trillion.[2] However, the specific number doesn't really matter. The key point is that *virtually everyone who will ever live has yet to be born.*

Well, with one massively important proviso, namely, that we don't go extinct. If we do, all that future value goes up in smoke. All that future flourishing, all that future exploration and discovery and beauty and meaning, all those trillions upon trillions of lives—gone.[3]

What a disaster that would be from the consequentialist perspective! Nearly everything of interest to the consequentialist *hasn't happened yet*. This, again, is simply a function of the distribution of human lives (and thus, potential sites of value/well-being) across time. There are 8 billion people alive right now. There are untold trillions of people yet to live, provided we don't screw it up. Even discounting the importance of future lives somewhat, so that we have *some* reason to prioritize present people over future ones, it still seems that the *future's moral value massively outweighs the present's.*

It follows that, for the consequentialist, it's *extremely* important to safeguard the future by reducing "existential risk," understood as any risk that "threatens the premature extinction of Earth-originating intelligent life or the permanent and drastic destruction of its potential for desirable future development."[4] In this vein, Bostrom proposes a moral rule called "Maxipok":

act so as "to maximize the probability of an OK outcome, where an OK outcome is any outcome that avoids existential catastrophe."[5] (One question for Chappell here: would he endorse Bostrom's Maxipok? If not, why not?)

Personally, I find the longtermist perspective just sketched quite compelling; if the argument is flawed, I don't see how. I'm inclined to agree, then, with Chappell that, since the "long-term future is immensely important," we have "extremely strong moral reasons to prefer better long-term outcomes."

And yet longtermism has some truly bizarre implications! Consider a thought experiment adapted from Torres:

> Before you sit two buttons. Push the first, and you'll save 1 billion actual people from being painfully killed tomorrow. Push the second, and you'll reduce the probability of extinction by some tiny amount. You can only push one. Which do you push?[6]

For the longtermist, the answer is clear, isn't it? You have to push the second button. Or so says Bostrom:

> The expected value of reducing existential risk by a mere *one millionth of one percentage point* is at least a hundred times the value of a million human lives ... Even the tiniest reduction of existential risk has an expected value greater than that of the definite provision of any 'ordinary' good, such as the direct benefit of saving 1 billion lives.[7]

If the reader is struggling to see why longtermism counsels hitting the second button, just remember how astronomically bad extinction is, from the consequentialist standpoint. Hundreds of trillions of lives are at stake! Nearly all of humanity's future hangs in the balance! Perhaps it will help to reframe the choice in explicitly numerical terms. Should you (1) save 1 billion people now, or (2) increase the probability that 500 trillion people will come into existence by some tiny percent? If your moral theory says, "maximize expected well-being," the math overwhelmingly favors option (2).

But wow, that's a tough pill to swallow. I should let a billion *actual* people die, just to reduce the risk of extinction from, say, 16% to 15.9999999%?[8] Torres concludes that we should reject any theory with such a crazy implication. I'm torn, because the logic generating this conclusion seems air-tight. I don't see any false premises. I'd love to hear Chappell's thoughts here.[9]

One final point in closing. I've focused on one type of existential risk, namely, the risk of extinction. But there are other ways humanity could fail to reach its potential besides literally dying out. Like a promising teen gone to seed, we might simply fail maximally to develop our capabilities. Bostrom calls such a scenario "permanent stagnation": "Humanity survives, but never reaches technological maturity—that is, the attainment of capabilities affording a level of economic productivity and control over nature that is close to the maximum that could feasibly be achieved."[10]

Or again, we might reach technological maturity, but in a "flawed" way, such that we realize "but a small part of the value that could otherwise have been created."[11] For instance, we might fail to develop cybernetics, or genetic engineering, or other "transhumanist" enhancements that would allow us to "explore hitherto inaccessible realms of value."[12] Imagine conquering aging and death; imagine radically upgrading our intellects, our capacities for joy and bliss, our artistic faculties, our bodies.[13] What a glorious transhumanist future we could have, filled with trillions and trillions of people smarter than Socrates, more aesthetically sensitive than Shakespeare, more athletic than LeBron James, more beautiful than Aphrodite, and on and on it goes. We could populate the galaxy with gods. Or, we could blow it—not necessarily by going extinct, although that's one way, but merely by failing to develop transhumanist tech. What a wild swing in utility these alternatives represent. Untold amounts of value hang in the balance.

In light of these remarks, we can run another version of Torres's thought experiment—one even more challenging, I think, for longtermism. Let the first button be as before; by pushing it, you can save 1 billion actual people from painful death. But the second no longer reduces risk of extinction. Instead, it reduces, by some tiny amount, the probability of a "flawed realization" as described in the previous paragraph. (Maybe by pushing it, you increase the odds of a genetic engineering breakthrough by .00001%.)

Which button should you push? Bostrom would say: the second. Many would take that as evidence against his theory. Again, I'm eager to hear Chappell's thoughts.

Notes

1. https://www.effectivealtruism.org/articles/fireside-chat-q-and-a-with-toby-ord
2. Bostrom 2013, 18.
3. Ord 2021, 44.
4. Bostrom 2013, 15.
5. Ibid., 19.
6. Torres 2023b.
7. Bostrom 2013, 19.
8. Ord 2021 estimates the probability of existential catastrophe this century at around 16% (or one in six).
9. Let me note, preemptively, that I don't think Chappell's "commonsense hybrid approach" to consequentialism helps him avoid the Bostromian verdict to hit the second button. Even granting that we have *weaker* reasons to bring new lives into existence than we do to "care especially about the well-being of antecedently existing individuals," I just don't see how that matters here. The math is just too overwhelming. There are just too many future lives at stake.
10. Bostrom 2013, 20.
11. Ibid.
12. Bostrom 2003.
13. Bostrom 2008.

13
BRENNAN'S RESPONSE TO CHAPPELL

Question 7: How Important Is Beneficence Compared to Other Values?

Chappell says:

> A good moral agent will prioritize morally more important ends over less important ones. This suggests that we can assess how morally good a person is (to a first approximation) simply by looking at how much they (want to) do to improve global well-being.

One way to think of morality (which I am not endorsing) is that there is something like a moral utility curve, a ranking of all possible states of affairs from better to worse. Whenever I act, I should try to generate states of affairs as high up on the curve as possible. I should choose the action which has the highest expected value. To this, one might then add that each person's welfare matters and matters the same. With a few more amendments, we're back to classic utilitarianism. As I argued in my section, of course utilitarians will endorse EA, though I think EA need not be utilitarian.

When Chappell asks about the importance of beneficence, we have in turn to ask what he means by importance. He could mean that beneficence trumps all other values; whenever beneficence and some principle (such as justice, non-maleficence, reciprocity, community, love, beauty, pleasure, courage) come into conflict, beneficence wins.

That seems implausible. Suppose two strangers are drowning to my left and one stranger to my right. I have one life raft, so I can't save them all. Sure, if nothing else that differentiates the people,[1] I agree I should save two over one. But suppose the two on the left are strangers and the one on the right is my wife. Here, I think it's not merely permissible, but required, that I save my wife.

Or suppose I find a wallet containing $500 cash. I wouldn't ponder how to use that money to maximize beneficent outcomes. Instead, I would return the wallet to the owner. Or, even if I could steal $50 from your bank account which you won't miss, and then donate the cash to Sightsavers, I wouldn't.

A sophisticated utilitarian might accept these conclusions. They might say what justifies the drowning case is that allowing people to form and act upon close personal bonds maximizes global utility. We allow people to make what are, in isolation, suboptimal decisions (saving one person over two) because the global effect of such rules permitted close bonds to outweigh the individual loses from such decisions. They might similarly argue that what justifies the wallet and bank cases is that having everyone adopt norms and institutions respecting property maximizes global utility, even though individual agents will make suboptimal decisions.[2]

To this end, consider an analogy from David Schmidtz. Imagine that at four-way stop signs, instead of first-to-arrive, first-to-go, we make it a rule that the neediest driver goes first. We'd have to leave our cars and talk it over to establish right of way; the rule would be too costly to implement. This rule wouldn't serve our needs. So, here is a case, a rule that tells us to ignore need better meets our needs.[3]

I think that's all correct, but I also doubt the justification of property rights, permissions to form special relationships, personal prerogative, and so on, simply reduce to asking what rules in general best promote welfare. Welfare matters, but it's not all that matters and it's not always the most important thing. For instance, many individuals are willing to suffer lousy lives to make great art or discover great mathematical knowledge; they do so because they regard these things as ends in themselves, not because they think the art or mathematics will eventually maximize welfare.

None of us live maximally beneficent lives and few of us feel particularly guilty about it. Peter Unger and Peter Singer say it's wrong to live high while people die. Though they give significant amounts to charity, they fall short of their demanding philosophies. Even if Singer donates 33% of his income,[4] his remaining post-tax income is enough to keep him in the top few million income-earners worldwide.[5] Unger and Singer might respond that this shows that they are morally defective people, but I suspect Ryan Davis (see his sections later) is right here—people don't feel guilty because, despite their statements to the contrary, they think this level of giving is supererogatory.

Here's an alternative view of morality, with a different view of how beneficence is important. Perhaps morality is not primarily about a ranking of states of affairs from better or worse or trying to maximize our place on a moral utility curve. Perhaps instead the default condition is that we may do as we please, but then morality imposes certain hard constraints on our actions, and then secondarily requires us to adopt certain general goals. On this view, the most actions are morally optional, neither forbidden nor required. We have some obligations (such as the duty to avoid murder, to respect others' speech rights, or to respect their rightful property) which remove various actions from our choice sets. We also have imperfect duties (to improve our character or capacities, to promote others' welfare, etc.), but retain significant discretion and freedom in deciding when and how to meet those goals. On this view, beneficence is indeed important, but it doesn't trump all other considerations or win every contest. This view allows that many other things are valuable and worth spending a life on.

Consider all the great achievements people have made in art, music, science, mathematics, sports, and the like. Consider values like friendship, community, and love. These things are part of what make life worth living. Without them, there would be little point in keeping people fed. But nevertheless, they are often worth pursuing even when their value does cash out in terms of welfare. We perform high musicals rather than feeding the poor not because we are selfish or callous, but because feeding the poor doesn't always win, even from a moral point of view.

Question 8: Does Self-Sacrifice Make Beneficence More Virtuous?

I agree with Chappell here. On the standard analysis, a virtuous person is disposed in context perform the right action for the right reason and feel the right way about it.[6] A person can do the right thing for the wrong reason. Imagine a sociopathic surgeon who saves lives because he wants money and fame. Her actions are good but her motives are not admirable. A person can do the wrong thing for the right reason. Imagine a person who wants to help but, thanks to negligent ignorance and misinformation, donates to bad charities. Her motives are good but her actions are not.

The point of beneficence is to help, not to sacrifice. Sacrificing is a means, not an end.

The point of beneficence is also not to *look helpful*. As David Schmidtz says, "If your goal is to show your heart is in the right place, your heart is not in the right place."[7] But, as I explained in my response to Question 6, we have strong evidence that most real-world giving is motivated more by the desire to promote the giver's status than to promote the welfare of others. Signaling is pervasive.

A truism in economics and management theory is that you get more of what you reward. A corollary is to be careful what you reward. For instance, I knew of a law school where the dean started giving $6000 bonuses for any academic publications, regardless of journal rank or quality. The policy reduced rather than promoted the school's research ranking; faculty stopped aiming high and instead churned out bad work in no-name journals.

Unfortunately, we tend to reward people for signaling virtue rather than exercising beneficence. We admire people who work in care professions, even when their marginal contributions are low, but frown upon those who work in logistics, even when their marginal contributions are high. We reward people for wearing colored ribbons and donating to high-profile charities, even when those charities have exhausted their room for funding and further contributions do little good. We ignore people who donate to low prestige but high output charities. Companies generate high social status by spouting platitudes. The typical social responsibility campaign starts and ends with anecdotes and pictures of smiling children rather than evidence-based cost-benefit analysis.

Sometimes, for these reasons, Effective Altruist communities develop their own pathologies. We are disposed to be conformist and to find markers of tribal identity which distinguish us from them. EA activists are not immune. This might explain, for instance, why polyamory (which has nothing to do with EA) is unusually prevalent among certain EA groups.[8] Many EA activists want to signal that they are more rational than others, and thus adopt expensive signals to prove their superiority.

Question 9: How Much Should We Care About Future Generations?

Suppose I leave a landmine under the street. It will probably kill someone, but I don't know whom. What I did was wrong.[9]

Suppose I place a timer on the landmine. It won't become armed and active for 200 years. I place it somewhere I expect people will travel, where it won't likely be removed or destroyed in the interim. Again, it will likely kill someone, but I don't know whom. In this case, it would kill someone who did not yet exist when I planted the bomb. Again, what I did was terribly wrong.

Suppose current birth-rate trends continue. If humanity can survive another 800,000 years, then approximately 100 trillion people are coming after us. If humanity can last for as long as Earth is habitable, there are approximately 40 quadrillion people yet to be born.[10] These people's lives matter and their welfare has to figure in some way into our decisions.

Economists generally think it's rational to place a discount rate on the future. Thanks to uncertainty, $100 real dollars today is worth more than $101 real dollars twenty years from now. ("Real" means I've corrected for inflation.) We can debate just how much, if at all, we should discount the welfare of future generations. But it can't be that we owe them nothing. Imagine I plant a bomb that will destroy all intelligent life (defined as life at least as intelligent as we are now) on Earth in 2 million years. Perhaps by then humanity will be extinct and no other species will be as intelligent, so the bomb will do nothing. Or perhaps it will instantaneously wipe out 20 billion people and stop quadrillions more from being born. Despite uncertainty about the future, planting that bomb would be evil, perhaps the worst thing anyone has ever done.

The harder problem is determining just what we owe future generations. Decision theory becomes sloppy when we have extensive uncertainty and arbitrarily large or infinite values at stake.[11] For instance, consider John Pollock's hypothetical "Ever Better Wine," wine which continually becomes tastier at a linear rate, without end, as it ages. Standard decision theory seems to suggest we should *never* drink it, because it is always better to wait until tomorrow. But if we never drink it, we never derive any utility from it.[12] Or, suppose there is a 0.0000000000000001% chance of extinction from an asteroid impact next year. Some decision-theoretic models suggest we should spend most of our money trying to avert that impact, since the expected disutility of such an impact is arbitrarily large. But many worry that this is a problem with relying on expected utility calculations in cases with arbitrarily large and small numbers and high degrees of uncertainty. It's easy to imagine countless horrible hypotheticals that could cause human extinction, each of which has a tiny but real chance of occurring. Should we be spending all our extra money fighting them all?

These are interesting debates, but regardless of how we resolve them, the implication is not that we may ignore future people's welfare.

Still, we should remember that future generations are likely to be significantly richer and significantly more technologically advanced than we are. Their capacity to solve problems will likely be far superior to ours. Consider: If world income continues to grow at a 2.5% rate (a conservative estimate), and given the UN's projection that world population will be about 11.2 billion in 2095, then the average person world-wide by 2095 will be as rich as the average German or Canadian right now. World product (in current US dollars) is on track to reach one or two *quadrillion* by 2100.[13] Compounding growth is a miracle.

So, one way to help the future is to solve problems now, if we can. Another way is to help ensure that the future is well-equipped to solve problems we lack the wealth, knowledge, or technological capacity to solve ourselves. Further, if future people are wealthier, this will resolve many of the on-going disasters of today which result from poverty.

Caring about the future thus means caring about economic growth. We can illustrate that further by considering how badly off we would be if there had been less growth. Suppose from 1930 until 2005, the US had had just one percentage point less economic growth per year (i.e., suppose it grew at 2.5% rather than 3.5%). In that case, by the 2005, US GDP would then have been *half* the size of what it actually was. Assuming the population is the same, that would make the US a middle-income rather than rich country. How much worse would our problems be? How much less ability would we have to deal with them? If, instead, growth had averaged 4.5%, we would be about 50% richer than we currently are. How much easier would it be for us to solve our current problems?

The Marxist philosopher G. A. Cohen once claimed that money—or rather the wealth it represents—is a form of positive liberty. Wealth is a ticket that helps you accomplish your goals.[14] So, one way we can help the future is to help them help themselves. That means we should continue to engage in capital accumulation and not sacrifice their welfare for ours, say, by stalling future growth in favor of short-term debt accumulation.

Notes

1 For example, imagine the two on the left are mass murderers and the one on the right is a scientist on the verge of curing malaria.
2 For example, Schmidtz 1995; 2006. Schmidtz is not a rule utilitarian per se but he makes arguments in this vein.
3 Schmidtz 2006, 31–32.
4 https://www.wsj.com/articles/peter-singer-on-the-ethics-of-philanthropy-1428083293#
5 https://howrichami.givingwhatwecan.org/how-rich-am-i
6 Hursthouse 1995.
7 I've heard him say this in person at least a dozen times.
8 https://time.com/6252617/effective-altruism-sexual-harassment/
9 Feinberg 1974.
10 https://80000hours.org/articles/future-generations/?utm_source=youtube&utm_campaign=2022-kurzgesagt-longtermism-link&utm_medium=youtube

11 For example, Arntzenius, Elga, and Hawthorne 2004.
12 Pollock 1983.
13 https://www.ubss.edu.au/articles/2022/july/what-will-the-world-economy-look-like-in-2100/
14 Cohen 1995, 58.

14
DAVIS'S RESPONSE TO CHAPPELL

It's very tempting to conflate what's morally good or virtuous with what other people will praise us for doing. As long we look like we're suffering for others' sake, it feels like we're doing really well as moral agents. We think that if we're fighting our reviled political enemies—especially when fellow partisans commend us for our solidarity—that we must be contributing meaningfully to our country.

Richard Yetter Chappell says that all of this is a mistake. However much we enjoy the praise of others, what matters morally is not doing things that are perceived as good, but doing what really is good. And however satisfying it might be to denounce political villains, this is not the best way of making a difference. I agree with all of this. There's a whole raft of biases that give us bad signals about what morally matters. As Chappell is well aware, we are constantly tempted to equate morality with currying favor with our in-group or fighting an out-group, but that pair of mistakes arises from a common source: confusing what matters to morality with what signals our allegiance to a cooperative group. As Chappell says, "Too often, our hearts belong to the culture war."

Chappell's essay offers an important corrective to a serious defect in our moral practice. But I can't follow him all the way. Where he thinks that "beneficence is, generally speaking, the most important value for guiding our moral lives," I prefer to say that beneficence is the most important value for guiding our altruistic activity. Of course, my variant is much weaker. The idea that beneficence is more important than everything else to our entire

DOI: 10.4324/9781003508069-18

moral lives is too rich for my blood. I like strong views in philosophy, so I admire the position as a philosophical commitment and even more as a real-life value. But I want to raise three questions about whether it can really be true. First, I will ask how revisionary his conception of moral virtue turns out to be. Second, I will ask how he sees the empirical facts: Can we revise our moral practices and retain what's good about our current relationships? Third, I will ask exactly what kind of personal disposition, on his view, counts as morally virtuous.

How Revisionary?

Chappell proposes assessing how morally virtuous a person is based on: (1) the strength of their beneficent desires, and (2) their contribution to promoting beneficent ends. He allows that his view may sound "somewhat revisionary." I want to consider his thesis that "if you want to be a better person, the place to focus is on trying to do more good, impartially considered." How revisionary, in principle, might it turn out to be?

First, let's consider this view with respect to one's own (non-moral) personal projects. Suppose I decide that I ought to be doing as much good, impartially considered, as I can. However, I am always succumbing to temptation—eating out with friends, splurging on baseball tickets, eyeing a summer vacation, and the like. In a cool moment, I realize that these desires compromise my beneficence. All of that money could have been donated to effective aid organizations. So I decide to adopt a regimen of practices aimed at purging these desires from my psychology. I try to systematically extinguish my love of watching the game, floating down the river, catching dinner with friends, etc., so that I won't succumb to my weak-willed inclination to spend the money on myself rather than give it away. Eventually, I more or less defeat my inclinations, and I look back from my new vantage point on my old self's passions as extravagant and tasteless. What should we make of this transformation? It seems to me that so far, this must be in keeping with the principle Chappell recommends.

Now suppose I realize that I could extend my efforts at self-reform to include activities I previously had thought were morally good. It's nice enough to attend a family reunion in the summer, but I have to admit that it would do more good, impartially considered, to donate the money I would have spent to effective causes. So that kind of thing should go, as well. Next, I realize that I could promote the impartial good even more if I persuaded others to abandon their self-directed desires, as well as most of their interpersonal altruistic desires. If only the high-powered lawyer would work more billable hours rather than make it home for her daughter's recital, she'd do better (assuming she donated the surplus money) at promoting impartial value!

I know that none of this is novel. Consequentialists and their opponents have been arguing over this sort of thing for more than a generation.[1] A common rejoinder to my question is that things just won't turn out this way: We will do better, impartially considered, by taking care of ourselves and the people we love. But now I worry that the Effective Altruist has inadvertently ruled that reply out. By showing how technological advances can turn cash into impartial value, hasn't Effective Altruist simultaneously made it more difficult to believe that we wouldn't really do better by thinking of our lives as warehouses of value to be optimally distributed to others?

The sharpest version of this question focuses on what Chappell refers to as "interpersonal warmth." Chappell takes for granted that interpersonal warmth in human relationships (being nice to friends, a considerate spouse, etc.) is easy to come by. From there, he argues that we should assess our moral lives on the metric of impartial goodness. But what if there's a tradeoff? Suppose I can best promote impartial moral value by belittling my friends and coworkers, haranguing them into giving their money away. Suppose I even realize I can do better yet if I violate commonsense morality, perhaps breaking promises to attend the recital or the reunion and giving the money away instead. In short, if I trade off the values associated with interpersonal warmth to secure impartial goodness, does that still make me a better person?

Now, I'm not saying any of this is true. J.C.C. Smart argued in that old debate over consequentialism and personal relationships that utilitarians are actually nice people.[2] And of course, he was right. Consequentialists can have friends, because they do! But the in-principle question still matters. Should we, in principle, be willing to sacrifice our projects and 'commonsense' moral commitments if doing so would promote impartial goodness?

What Would Really Happen?

Let's next imagine that people actually lived by the criteria for personal virtue that Chappell sets out. What would happen? Such a world would involve reorganizing moral priorities from the values associated with "interpersonal warmth" to the value of impartial beneficence. If I read him correctly, Chappell supposes that we don't have to worry too much about the tradeoff. After all, "there's no general lack" of interpersonal warmth, and there's a great deficit of beneficent agents. However, I'm not sure we can be confident. There's no lack of interpersonal warmth in the actual world, but the actual world is also a place where people put a moral premium on that value. If people valued warmth toward others much less in comparison to other values, how could we be sure that there would still be enough of it to go around?

Chappell might regard this worry as overblown. After all, aren't there lots of Effective Altruists in the actual world, and aren't they (at least very often) nice people? I think there are a few reasons for caution. First, some ethnographic work on individuals in the Effective Altruism movement goes into detail about how committing to impartial beneficence can pose a real challenge. Larissa McFarquhar's *Drowning Strangers* tells a series of stories in which a commitment to beneficence can make a person's life or relationships really difficult.[3] To offer just one story, Julia and Jeff (a young couple) committed to give their money to effective causes, and then quickly found that their values could be a constant source of guilt and recrimination. They couldn't so much as buy a single candied apple without the thought of the family deprived thereby of some life-saving measure. This is a hard life to keep up.

It's true that many Effective Altruists are not so afflicted—and also that members of the movement eschew blame toward others. But it doesn't follow that if other people *were to become* Effective Altruists, their experiences would be similarly positive. In general, it's extremely perilous to make inferences from any self-selecting sample to a broader population. I suspect that there's something about currently practicing Effective Altruists that makes their lives different from what others would experience if they lived by the same principles. When altruism works well, it does make the altruistic person's life better. The way that happens is through the altruistic person's sense of autonomy—their control over their own life. The connection between an individual's intrinsic motivation, or their sense of self-determination, is what makes all the difference.[4] If you're an Effective Altruist right now, you're probably intrinsically motivated by its values. But doing something because you feel constrained by morality does not promise the same sense of self-determination. There's a whole raft of ways in which altruism can be bad for the person doing it—what psychologists call "pathological altruism." My worry is that Effective Altruism is meaningful within the lives of those choosing it right now precisely because they are choosing it right now. We can't make inferences about how their values would fit for others. But very generally, human values are highly heterogeneous, and it's difficult to durably expand one's identity to incorporate new values—let alone give up many of one's current values. That's just not something humans can do very well. My question for Chappell is: What do you think a world in which people who lived by a norm of impartial beneficence would be like for the people living by that norm?

Which Disposition?

My final question is about the virtuous agent. As noted above, Chappell's argument is that we shouldn't worry about a bunch of moral side-quests: defeating enemies, collecting praise, and the like. Instead, we should just focus on what is good for others. Then Chappell says, "The truly virtuous agent simply cares about helping others, and will take whatever permissible means best serve this end." But now it seems to me that this claim about virtue must presuppose a premise of the form: The agent who simply cares about helping others will do the best at promoting impartial value. However, it seems like this premise might be false. Just as an agent might do better at promoting impartial value by disregarding the value of moral praise, isn't it also possible that an agent might do better at promoting impartial value by not caring about helping others? Instead, wouldn't it be more accurate to say that on this view, the truly virtuous agent will have whatever disposition is such that having that disposition will cause them to promote the impartial good as much as possible?

To see the idea, imagine a character who cares a lot about helping people, and that caring makes her help whoever she happens to encounter in the course of her day. But as it happens, most of those people are already relatively prosperous and healthy. Helping them is inefficient, and our protagonist realizes that if only she could get herself to care less about helping people and more about—say—making money and almost-nothing-else, the more good she could do. Her disposition would be less directed toward helping others, but more optimal from an impartial standpoint. Is she thereby more virtuous?

Conclusion

I've sketched a few questions that share a common pressure point: Isn't Chappell's revision to our conception of personal virtue somewhat more revisionary than we might have initially thought? You might think I am pushing a skeptical line about his view. But, to be clear, that's not right. Conditional on there being some moral facts at all, it's true that I think it's most likely that morality takes the form of the kind of rights-based minimalism I describe in my contribution to this volume. But the second most likely way I think morality could be is roughly as the utilitarian sees it. The views I find implausible are those based on ad hoc first-order intuitions about cases. So my aim is not so much to oppose Chappell's view as to call attention to its radicalism. I'm not sure whether he'll agree in his answers, but I think that the truth about morality—if there is such a truth—will turn out to demand a lot of changes from our ordinary moral intuitions.

Notes

1 For illustrative examples, see Scheffler 2012; Jackson 1991.
2 Smart and Williams 1973.
3 MacFarquhar 2016.
4 Weinstein and Ryan 2010; Martela and Ryan 2016; Deci and Ryan 2014, 3; Ryan et al. 2021.

15
CHAPPELL'S RESPONSE TO ARNOLD, BRENNAN, AND DAVIS

Thanks to my interlocutors for their stimulating thoughts! Below, I'll revisit my three themes of *beneficence*, *virtue*, and *longtermism* in light of their probing commentary.

How Important Is Beneficence?

Arnold and Brennan both stress that beneficence does not intuitively *trump* all other virtues and values. Arnold, for example, writes that the purely beneficent agent is "a deeply alien and unappealing character, someone whom basically no one outside a philosophy seminar room would ever want to be, or befriend, or marry." But I wonder how much weight this observation really carries. After all, we have obvious self-interested reasons to prefer to associate with *reliable allies:* those who would favor us over others, regardless of what morality actually demands. (A slave-owner wouldn't want to marry an abolitionist, but that hardly shows that the abolitionist is morally mistaken.) We also have obvious self-interested reasons to prefer moral ideologies that endorse the actions we independently *want* to perform. (Few people want to be vegan, but it hardly follows that our abuse of factory-farmed animals is morally acceptable.)

So, it could just be that morality asks things of us that we aren't willing to give. Since we also don't want to think of ourselves as morally sub-par, we make up self-serving stories about how it's actually totally fine to drink champagne when we could instead be saving children's lives. But our preference for self-serving ideologies is no reason to think that they're actually correct.

While I find this debunking explanation plausible, I didn't actually mean for my starting essay to be *that* controversial. As I wrote: "Most people presumably meet the minimal standards for *not being evil*; what's left unsettled is just *how good* we're going to be." For someone who is morally typical (featuring many commonsense virtues and a strong respect for others' rights), the greatest *marginal* impact on their virtue would come from further increasing beneficence. There's more moral upside to above-average beneficence than to above-average levels of other virtues. That modest claim is compatible with holding that extra beneficence cannot compensate for outright *viciousness* like torturing the innocent. The marginal moral value of different virtues may vary depending on their current level of development.

Davis raises the interesting worry that one way to increase one's beneficent impact may be to deliberately *purge* one's competing motivations—from hobbies to familial love. This probably isn't advisable in practice: EAs warn against "burnout" from becoming too hard-core or monomaniacal in one's pursuit of the impartial good. But let's make things difficult by supposing that a magic motivation pill could keep you going, however bleak your life became. What then?

I think there's a difficult tradeoff here between different kinds of normative considerations. In purging your interests in interesting things, and your love of lovable people, you would be making yourself more irrational: deliberately blinding yourself to genuine reasons, and cutting yourself off from genuine goods. But it would be in the service of acting upon other reasons (of beneficence) that are objectively more weighty, though we glimpse them only faintly, and hence ordinarily fail to be sufficiently motivated by them.

So, yes, the sacrifice Davis describes could potentially be worth it. But it is not without cost. When we only faintly glimpse the strongest moral reasons, moral outcomes may be achieved via the distasteful means of "leveling down" our grip on all competing reasons. (The same may be true even on non-consequentialism, if you imagine someone who has very weak motivations to respect others' rights, and these weak motivations can win out only if the agent purges themselves of all competing motivations.) Just as the non-consequentialist would prefer to instead solve their moral mismatch by sufficiently *increasing* the agent's respect for rights, so consequentialists should prefer to solve *our* mismatch by sufficiently *increasing* the agent's beneficent concern for others.[1] That way, they could successfully act on their strongest moral reasons without having to blind themselves to other genuine reasons.

Given this general picture, it will come as no surprise that I disagree with Brennan when he writes: "We perform high school musicals rather than feeding the poor not because we are selfish or callous, but because feeding the poor doesn't always win, even from a moral point of view." I rather think that we enjoy these other activities only because we have managed to ignore the plight of the poor. If I do something that I truly think is more morally important than helping the poor—donating to an *even more* cost-effective animal welfare charity, for example—I could justify my choice to those I overlook. But how many of us could honestly *justify* to a starving child our choice to prioritize the cultural enrichment of our own children? How many would still enjoy the musical if the backdrop included a screen showing the starving children we passed over in order to fund the show? I expect most people would feel massive guilt in such a circumstance (and probably anger in reaction to the unwelcome guilt-tripping), revealing that we don't *really* believe Brennan's claim that our choice was fully justified, however much we'd *like* to believe it.

Let's wrap up this section by revisiting the challenge of the wicked billionaire philanthropist. Arnold imagines a variation of the case in which the torture isn't malicious; rather, it is (somehow) necessary to rejuvenate the agent and enable them to achieve greater good the next day. Maybe they're a vampire. (It would probably still take an unusually callous person to be willing to torture an innocent person with their own hands, even if this generated greater benefits for others; but let's suppose our vampire has an alien psychology, such that they find the wails of the distant needy just as salient as those of their immediate victims.)

While many won't like this conclusion, I think the vampire may well be justified in this case. To see why, suppose first that the vampire only drains a subset of those who would otherwise die of malaria without his aid. And suppose you ask the potential malaria victims whether they'd rather die of malaria or take their chances with the vampire. Presumably they'd choose the vampire (this is to their expected benefit, given the stipulation that the vampire does more good than harm to this population). To empower the malaria victims to make this choice, it seems we must take rights to be alienable: the malaria victims may waive their rights against vampire attacks, in order to receive the greater benefit of protection against malaria. (Who are *you* to tell the malaria victims that they must die rather than allow the vampire to treat them in a way that offends your rarefied sensibilities?) Since the vampire is acting in accordance with the *ex ante* unanimous rational preferences of all affected, their action cannot be wrong.[2]

Next suppose that the vampire chooses victims randomly: they don't exclusively prey on those vulnerable to malaria. This is more fair than what we previously imagined. And a moral improvement upon a permissible act must itself be permissible. Suppose the vampire picks me, and I complain that I never waived *my* rights. But, he reasonably points out, that's just because I was in a position of unjustified privilege to begin with. There's no moral reason why others should be susceptible to malaria when I am spared. From behind a "veil of ignorance," not knowing my privileged position in society, I would agree to the same deal that the malaria victims wanted: to waive my rights, conditional on others doing likewise, whenever this would be for the greater good. This would best serve my *ex ante* interests, from behind the veil.

Yet now the veil is lifted, and—seeing my privileged position—I suddenly demand my rights back, never mind those dying from malaria. Isn't that *transparently* special pleading (like reneging on a fair lottery only *after* drawing the short straw)? It would clearly be more ethical to stick to the agreements we all would have made from behind the veil. And so I cannot complain about the vampire's making a meal of me, given that this truly is a necessary means for securing greater benefits for others: benefits that I myself would have wanted, and agreed to, had I an equal chance of ending up in their position.

Virtue and Pathological Altruism

Davis wonders whether the virtuous nature of benevolence depends on empirical contingencies: "isn't it ... possible that an agent might do better at promoting impartial value by not caring about helping others?"

This is possible, but it wouldn't change what's virtuous. It's just to say that we can imagine situations in which virtue would be morally unfortunate, much as we can imagine cases in which instrumental rationality is instrumentally unfortunate for an agent.[3] I take virtue, like rationality, to be an *intrinsic* feature of an agent: any intrinsic duplicate of a virtuous person is also a virtuous person, even if their circumstances vary such that their virtue turns out (bizarrely) to have bad consequences.

Suppose that an evil demon likes torturing puppies, but he likes *even more* to do the opposite of what Bob wants. So in fact some nearby puppies will be spared if and only if Bob wants the demon to torture them. Now suppose that Bob notices the puppies and cruelly intends to bring them to the demon's attention, hoping that the demon will torture them. (He doesn't realize that this will actually have the opposite effect.)

Now let's ask: are Bob's intentions "good"? There are two very different ideas we need to distinguish here. Bob's intentions are certainly *morally fortunate*—they're the intentions an informed benevolent observer would want him to have, since they will serve to save the puppies, and thereby promote the good. But, equally obviously, Bob's malicious intentions are not "good" in the sense we normally have in mind when speaking of "good intentions": Bob does not *mean well*. He does not intend or aim at anything good (quite the opposite). So these malicious intentions—however fortunate they turn out to be—do not reflect well on Bob. Since what Bob intends is bad rather than good, there is an obvious sense in which he has "bad intentions," and is messed up as a moral agent. This is to assess his intentions in terms of their intrinsic *warrant* or *fit* with the moral truth, in contrast to our earlier evaluation of them in terms of *promoting value*.

In this scenario, we should *want* Bob to make a moral mistake, because that has better results. But we should not lose sight of the fact that Bob is indeed *mistaken* in his malicious desires. The latter verdict is reflected in our judgment that malice is a *vice*, however fortunate it may turn out to be in the circumstances.

Someone who truly wants the best for everyone is a good person, even if they fail in their worthy goals. Their virtue may be morally unfortunate. But it is admirable, nonetheless. What's admirable and what's preferable may come apart in such cases, since the former concerns the target's intrinsic features whereas the latter additionally takes extrinsic considerations into account.[4]

Davis additionally worries about "what would *actually* happen" if more people became convinced of effective altruism: maybe effective altruism works well enough for those who have self-selected into it, but would prove detrimental if others felt *pressured* into it. While anything's possible, I'd be surprised if greater generosity on the margins didn't have good overall effects. Maybe there's a point at which greater altruism would become counterproductive, but I think most of us are currently a long way short of that point.

But I also think the question is underspecified. What would happen if people felt *pressured* into EA may be very different from (and worse than) what would happen if more people became convinced of effective altruism as a result of feeling appropriately *enticed* by the excellent reasons for supporting the project. I strongly prefer the latter prospect to the former. Even if it is an obligation (a question I remain neutral on), it's generally a sad thing if people have to be pressured by an alienating sense of duty into fulfilling their obligations. The happier scenario is for people to inculcate virtuous motivations that lead them to *whole-heartedly* do as they ought.

A world in which *everyone* possessed the virtue of impartial benevolence to a non-trivial degree (supplementing their various local and particular concerns) sounds pretty utopian to me. But the more practically relevant question is whether it would be good for the world to have marginally *more* impartial benevolence than it currently does. And the desirability of *that* seems, if anything, even clearer.

Longtermism

Arnold and Brennan both agree that we clearly should take future generations into account. But they raise the tricky question of whether we should allow the astronomical stakes of the far future to completely swamp present-day interests. Should we, for example, prioritize some tiny reduction in extinction risk (say, a mere 1 in a billion permanent reduction to the absolute level of risk) over saving a billion current lives for certain?

This is a tricky issue to think about.[5] But we might work toward a tentatively justified conclusion by starting with some easier questions.

First, we can ask whether one can simply *ignore tiny probabilities*. Here the answer is very clear: No. To see why, suppose that:

1. a killer asteroid is on track to wipe out all life on Earth, and
2. a billion distinct moderately-costly actions could each independently reduce the risk of extinction by 1 in a billion.[6]

Now, it would clearly be *well* worth saving the world at moderate cost to each of a billion individuals. And, since each of the relevant actions is *independent* of the others, if all the billion actions are collectively worth performing, then any one is equally worthwhile no matter how many others are performed. So, it is extremely worthwhile to oneself perform one of the risk-mitigating actions, despite the moderate cost and tiny probability of making a difference.[7]

Next, suppose that the costs of each action become extreme. To reduce the risk of extinction by 1 in a billion, an individual must now sacrifice their own life (when they could otherwise live several decades before the asteroid hits). Such heroism may be "too much to ask," in a sense. But it would not be *pointless* self-sacrifice: it would very much be *worth it* for the billion saints or heroes to choose to save the world at great cost to themselves. It wouldn't be like sacrificing themselves merely to spare someone else a papercut, in which the cost outweighs the benefit. Saving the world really is a greater benefit than the cost of a billion deaths. And if that's so, then an independent 1-in-a-billion permanent reduction of extinction risk is a greater benefit than the cost of one painful death.

Next, suppose that more contributions are required. Suppose there are somehow a billion billion (i.e., one quintillion) more people on Earth than we'd previously realized. And one quintillion distinct extreme sacrifices could each independently reduce the risk of extinction by 1 in a quintillion. Again—assuming the potential future lives would be sufficiently good and numerous—this would clearly be worth it for all quintillion to do, and hence (by independence) for any number of them to do regardless of how many others do likewise. So, in particular, it would be worth it for one billion saints to heroically sacrifice their lives in order to (collectively) reduce the risk of extinction by 1 in a billion.

You shouldn't try to stop the saints from making this worthwhile sacrifice. So we find that permanently reducing extinction risk by 1 in a billion is indeed better than saving a billion lives (given suitable assumptions about the potential future). This isn't a "tough bullet to bite": we can *see* that it's right, just by thinking through the above series of cases.[8]

Crucially, this argument assumes that we *know* the correct probabilities that apply in the scenario. In real life, that's unlikely to be the case. Anything so speculative that you're inclined to assign a "1 in a billion" chance to it might objectively warrant a vastly lower credence. It might also have a comparable chance of proving *counter*productive, or making the prospects for humanity worse rather than better.[9] In such a case, my above argument does *not* commit us to prioritizing baseless speculation over reliably saving lives. The intuition that we should ignore *baseless speculation* may well be correct. But that's not the same as dismissing *tiny probabilities*, since it is at least possible to imagine cases in which those probability assignments are robustly well-grounded.

So, I agree with Brennan that we can't rely on naïve expected-value calculations given "extensive uncertainty" about the probabilities themselves. Those calculations are *garbage in-garbage out*: they only work with accurate numerical inputs. So the hard epistemic work remains to determine which speculative threats we should assign non-trivial credence to, and which interventions are robustly positive in prospect. Those aren't things that I can answer. But the stakes suggest that the questions are worth taking seriously. To mistakenly dismiss a serious x-risk would be one of the worst things we could possibly do. Worse, even, than passing up a sure opportunity to save a billion lives.

Notes

1 I call this the 'leveling-up' interpretation of utilitarian impartiality, in contrast to the distasteful 'leveling down' that Davis imagines: available at: https://www.goodthoughts.blog/p/level-up-impartiality
2 Hare 2016.
3 See the discussion of "rational irrationality" in Parfit 1984.
4 I expand on these two dimensions of moral assessment in Chappell forthcoming.
5 Fortunately (for our ability to understand it), the expected value of the future is still only *finite*, and not *arbitrarily* large, so we avoid some of the decision-theoretic paradoxes that Brennan flags as a concern.
6 A similar case is also discussed by Kosonen 2023.
7 We can generalize the case to make the probability arbitrarily small, so long as the stakes are correspondingly increased. For arbitrarily large X, suppose that a population of X individuals is threatened with extinction, and each individual has the opportunity to independently reduce the risk of mass extinction by 1/X. This action is clearly worth taking (for all, and hence for each). So we find that probabilities as tiny as 1/X can be worth taking into account, no matter what number you put in for X.
8 If you doubt our starting assumption that extinction is extremely bad, just replace it with the risk of a permanent dystopia (since nobody can reasonably doubt that massive suffering is extremely bad). It's trivial to modify my argument to demonstrate that reducing the risk of *permanent dystopia* by 0.000000001 is clearly better than saving a billion lives with certainty.
9 Hiller and Hasan forthcoming.

PART IV
Limiting Beneficence

Ryan W. Davis

16
LIMITING BENEFICENCE

Ryan W. Davis

No one but you has any rights to your physical person. Others can ask for your help, and perhaps sometimes even take resources from you. But your physical body is morally off limits to anyone else. If others have no claims to our physical bodies, then what we do with our bodies is up to us. My thesis is that there is no moral requirement to provide aid to others. Choosing to provide aid to others—like all positive actions—is a matter of individual moral discretion.

Effective Altruists exercise their moral discretion to sacrifice for the good of strangers. They go beyond what duty requires, and so their project is mostly free from moral risk. It's good to do good.

If Effective Altruism is morally admirable but not morally required, should you do it? Positive psychology connects altruism with well-being. Happy people help others, and people who help others are happier.[1] But the trick is that most of those happiness-making good deeds are also voluntary. Choosing to help others makes people feel like they're making a difference as agents. It makes them feel autonomous.[2] The people for whom altruistic behavior is good are people who care about choosing to live that way. Whether you're one of those people is for you to decide.

Question 10: Is There a Moral Obligation to Beneficence?

I believe there is no moral obligation to beneficence. Providing aid to others in need is a great moral good. However, the fact

DOI: 10.4324/9781003508069-21

that it would be morally good to help those in need does not give them any special rights over you. This means that no matter how important or urgent it might be that others receive aid, and no matter how well positioned you might be to provide it, you don't morally have to do it. You are not guilty of any wrongdoing if you do nothing.

In this section I will try to persuade you of two things. First, there's no moral obligation to help other people. And second—crazy as it sounds—I'll try to persuade you that you probably *already believe* there's no moral obligation to help people.

Let's start with two simple arguments. First, imagine you are swimming at the beach, and—out of nowhere—a rip tide pulls you out from the shore. Happily, a stranger who just happened to be passing by jumps into the water, swims out to you, and pulls you to safety. What attitude would you feel toward the stranger? When I think of the stranger saving me from a watery death, I imagine feeling deeply grateful. Next, think about the concept of gratitude. A standard view in conceptual analyses of gratitude is that gratitude is warranted to a person only if their action has exceeded what they were morally required to do. I might say "thank you" to someone who gives me the correct change, but that kind of politeness is different from gratitude. Gratitude is directed to another person, and it communicates to its recipient the sense that their action has created a kind of moral imbalance.[3] If I feel grateful to you, I acknowledge that you have done something for me that I could not have demanded. The empirical literature finds that this feeling is common; people associate gratitude with being in another's debt.[4] And it makes no sense to feel oneself in another's debt if they did only what they were morally required to do. In that case, they would have done only what was owed to you already, and so you would not accrue any new debt to them. If my gratitude to the stranger was warranted (as I believe it was), then they were not morally required to save me from drowning. And if there isn't an obligation to provide aid in the extreme case—when another's life is on the line—then there also is no obligation to provide aid in less extreme cases, such as when one could provide a benefit to another. I conclude that there is no obligation to beneficence in any case (or, at least without actively accepting such an obligation). We can put the argument this way:

1. Gratitude is warranted to a stranger who saves you from drowning.
2. If gratitude is warranted for some action, then that action was not morally required.
3. So, a stranger is not morally required to save you from drowning.
4. If a stranger is not morally obligated to save you from drowning, then there is no obligation to beneficence.
5. So, there is no obligation to beneficence.

Call this the *argument from gratitude*.[5] Skeptics will often deny (2), holding that gratitude might convey only that an action was very difficult to perform, or that most people wouldn't have done it. In that case, it might make sense to feel grateful for actions that are also morally required. I am not persuaded. Imagine a person possessed by such a powerful desire to trip you as you pass them on the way to your seat that they can only restrain themselves with great difficulty. (Just to make the case vivid, let's suppose you're at a baseball game wearing Astros gear, and the sight of Astros fans fills them with lingering rage over your team's past cheating scandal.) Now, let's suppose they comply with their obligation to not harm you only by great effort. Still, you don't have to feel grateful to them. Likewise, even when most people are violating the rights of others, you don't have to feel grateful to someone who merely *withholds* from violating your rights. In my view, the most coherent account of gratitude contrasts being grateful with mere moral elevation—or recognizing a morally rare or excellent feature of another agent's action. I think gratitude gives us an important clue about the limits of what we can morally demand of others, and so too what they can demand of us.

Second argument. Imagine you are walking along a dock when you see a child drowning in a deep lake. You are unable to swim, and so you cannot save them yourself. You also happen to see someone else strolling the dock whom you know to be a strong swimmer—strong enough that they could save the child without incurring any danger to themselves. You call out to them, pointing to the child and pleading with them to jump to the rescue. Strangely, they refuse your petition. Let's suppose you know that if they were forced into the water, they would then save the child. They're just unwilling to get in. This gives you an idea. You have a gun in your backpack, and you threaten them with it unless they jump into the water. Is it permissible to threaten the capable swimmer standing next to you?

My own view is that threatening the capable swimmer is wrong. The fact that they refuse to come to another's aid does not make them liable to any interference—either by brandishing a weapon or in any other way. For example, it wouldn't be any better to grab them and throw them off the dock, even if it would not injure them and even if there is nothing else you could do. There is no in-principle way of inferring that would be morally permissible. People have rights against physical force. With that central idea, we can set up the following argument:

1. If someone is morally required to perform some action, then it is in-principle permissible to interfere with them in order to bring about their performance.
2. It is not morally permissible to interfere with someone in order to bring about that they come to the aid of strangers.
3. So, aid to strangers is not morally required.

Call this the *argument from noninterference*. I've said something about the intuition supporting (2). Defenders of duties of beneficence will probably deny (1). Instead, they will argue that we can be subject to moral requirements, even if those requirements cannot be permissibly enforced. Here I want to distinguish between two different conceptions of moral requirement. One way to think of moral requirement is in terms of what one has most (or, decisive) moral reason to do. A second way of thinking about moral requirement is in terms of what another person is entitled to demand of us, to resent us for not doing. My use of the concept 'moral requirement' follows the second approach.

To see the difference, consider that we may sometimes have decisive reason to perform some action, where that action is morally good. However, it may be that no one is entitled to our action. It might be irrational for us to fail to do it, but no one could resent us. I think cases of this kind are very common. You could help a friend move or take a neighbor to the airport. It might be that you have nothing else going on and wouldn't mind doing it. (Let's say, then, that you have decisive reason to do it, and it would be a morally good thing to do.) Doing something else would be a mistake.[6] But they couldn't appropriately demand that you do it. You don't owe to it to them to help.[7]

Why care so much about this restricted notion of wrongdoing? The answer is that morality matters to us because it provides rules in our interactions with other persons that are non-optional—that in some sense we must follow. No other normative domain is like this. You can say you just don't care about being a good sports fan, or a good student, or about even about following the rules of etiquette. Morality is different. We cannot sluff off morality's demands by insisting that we don't care about it. Morality is categorical.

Now let's go back to premise (1) in the argument from noninterference. Why think that if an action is morally required, then it is permissible, in principle, to interfere with someone for not doing it?[8] We've now added the idea of a conceptual tie between moral requirement and the standing of another person to demand that we perform the action, or to resent us if we don't perform it. To say that an action is required is to say that it is someone else's business. And not just one other person's—many people could blame you for not doing what you were morally required to do. Of course, you could insist that while another person does have standing to blame you if you don't do it, they would not have had standing to interfere with you to make you do it. But this seems like a delicate needle to thread. The action would have to be someone else's business after the fact, but no one else's business before the fact. And that seems unstable. As Kieren Setiya writes of wrongful actions, "If a stranger could have compelled you not to perform the act, beforehand, that would not have infringed your rights; blame is just the residue of that fact."[9] The idea, in short, is that if moral requirement is not conceptually tied to the possibility for interference, then there is no functional profile for the concept at all.

We now have two arguments for the conclusion that acts of beneficence are not morally required. Yet this claim seems to depart from powerful human intuitions. As Effective Altruists have long pointed out, if you see a child drowning in shallow pond, it seems horrific not to save them just to avoid ruining your expensive shoes. This intuition seems so obvious that Effective Altruists have largely taken it for granted.

So what should we say about the intuition that we ought to save the drowning child? I think the intuition is getting at something which is importantly correct. It does seem obvious that we ought to save the drowning child, that this is *the thing to do*, and that—in the circumstances—we would do it. But none of that is the same as saying it's morally required, at least as I've understood that idea here. So now the question is: When people affirm the intuition that they ought to save the drowning child, how should we interpret their report? I can't settle this question here, and it's hard to test directly. (Just try to get IRB approval for a psychology experiment in which you push children into shallow ponds!) Still, I think there are several reasons to believe that most people don't accept a moral requirement to aid.

To start, a first thing to notice is that it makes no difference whether you tell people that Effective Altruist giving would be a good thing to do, or that it would save lives, or that they're required to do it. Varying that language in the arguments you give people to engage in giving to aid organizations makes no difference.[10] It seems reasonable to think most people are not sensitive to it. The real questions are: How do they act and how do they expect others to act? And on these questions, people are remarkably unlikely to engage in giving, are unlikely to expect it of others, and are even unlikely to praise people for it. Think about it this way. Imagine you learned that a close friend of yours had secretly embezzled money from their employer—a local school—for many years, and that the whole time they had lied to everyone about it, including you. My guess is that learning this fact might well change your relationship with your friend. You might be inclined to blame them, to trust them less, and maybe even to end your relationship. In short, you would treat them like they had done something morally wrong.

Now imagine learning that your friend did not give any of their money away to charitable organizations. However, they also never lied about it, and they didn't take anyone else's money that they didn't have a right to. My guess is you would not think of this as any reason to criticize your friend, blame them, or change your relationship with them. In short, you would not treat them like they had done anything morally wrong.

Here is my explanation: We don't really think that failing to aid others is morally wrong. It's just socially undesirable to admit this out loud. There are a host of issues that work the same way. People will say whatever is socially desirable to say on a variety of topics, but that doesn't mean they believe it. For example, political psychologists find that people report caring a great deal about social equality, democracy, and the give and take of ideas.[11] But when you dig just very slightly deeper, it turns out that people will overwhelmingly give up on all of those values for the sake of getting the outcomes they favor.[12] So it is not surprising at all that people will report the intuition that they ought to save a drowning child. But socially desirable reports are a notoriously poor proxy for real beliefs. People's own actions don't support the view that they really believe aid is morally required.[13] Neither do their blaming attitudes toward others.[14] And neither do their expectations about when other people will provide help to them.[15]

It might be surprising to think that we have no obligations to help other people. But not only is this true, I believe, it's what most of us have really thought all along.

Question 11: Is Effective Altruism Morally Risky?

Some philosophers have worried that Effective Altruism carries moral risks. They fear that Effective Altruism will run afoul of some other unconditional obligation, or else that Effective Altruists take on conditional obligations, which they might then fail to fulfill. I don't share any of these worries. I think that Effective Altruists are doing something morally extraordinary. The central moral risk to avoid is in placing unwarranted moral demands on others. Effective Altruists should not do that.

Unconditional moral worries include questions like: Is Effective Altruism undemocratic? Does Effective Altruism underprioritize the right moral values? Does Effective Altruism promote a bias in favor of the institutional status quo?

I'll consider these in turn. First, consider the accusation that Effective Altruism is undemocratic. Democratic institutions are supposed to be transparent: Their workings are (usually) meant to be accessible to citizens. Democratic representatives can (in principle, at least) be held accountable by their constituents. Those who are dissatisfied with their performance can campaign against them. Private philanthropic foundations, by contrast, don't have to be transparent, and cannot be held accountable by their constituents—at least not electorally. As Robert Reich puts the point, "there is no way to vote out" the Gates Foundation.[16]

Second, consider the worry about the right values. Some philosophers have worried that Effective Altruism might be unduly influenced by utilitarianism. Some have argued that this might lead the Effective Altruist into moral error. For example, Effective Altruists might favor a policy that promotes utility over one that prioritizes the least well off (if, by hypothesis, aid given to members of the latter group would do less to promote utility overall). Or, Effective Altruists might allow for the continuation of rights violations (say, through sweatshop labor) if allowing these violations were welfare-promoting on the whole.[17]

Third, consider the fear that Effective Altruists shore up the status quo. The idea is that Effective Altruism relies on global capitalism to acquire and disseminate resources. As a result, it engages in the "subordination of benevolence to the market" and thereby helps to "stabilize the system that results in poverty, disease, and environmental destruction."[18]

My response to these worries will be straightforward. On my view, the Effective Altruist is doing something morally good, but not something anyone could morally demand. And if you are doing something morally supererogatory, then there is no pressure to think that you ought to be subject to political transparency, democratic accountability, or that you ought to promote rights-satisfaction or economic transformation. Think about ordinary, first-personal cases. Suppose you decide to spend your

Saturday volunteering at a school fundraiser. You need not allow other parents to vote on how many times you volunteer, or which school you contribute to. You don't have to be transparent—showing other volunteers your Google calendar for the week. You are already doing more than you morally have to do, so no one has any further claim on you.

The same goes for the worries about utilitarianism. Perhaps Effective Altruists—by and large—prioritize utility to rights. If they are doing more than what's required anyway, the fact that some rights are going unsatisfied is not their fault. No one has any claim against them that they do more to promote rights. Likewise for economic transformation. Some suppose capitalism to be a great moral evil. That is a controversial view.[19] Right or wrong, they have no standing to blame Effective Altruists more than any of the rest of us who live our lives within a capitalist economy. If anything, they should still think Effective Altruism is morally good: At least the Effective Altruist is concerned about the same moral problems that the opponent of global capitalism purports to care about.

So, I am not motivated by the thought that Effective Altruists are violating some unconditional moral requirement. But what about the idea that in participating in Effective Altruism, they thereby adopt some further moral requirement (not shared by everyone), and that this creates a new kind of moral risk?

It will help to sharpen this idea with a case. It's common enough to think that sometimes we are not morally obligated to perform some action, but that if we choose to do it, then we bring ourselves under a new set of moral requirements with respect to how we do it. Imagine, for example, deciding to buy a birthday present for a friend. You might think you don't have to buy a present at all, but if you do get one, you ought to get a present your friend would like, rather than something you'd personally enjoy buying. A more philosophically minded case: You dash into a burning house to save a child. By hypothesis, you aren't required to do this. (After all, you could die!) But when you're inside, you discover a second child whom you could also save without incurring any additional risk to yourself. It's common to think that in this case, you're morally required to save the second child.[20]

Perhaps the Effective Altruist is like the moral hero who runs into the burning house. Once inside the house, the argument goes, you have to do as much good as you can. Likewise, one might think that once you've committed to be an Effective Altruist, you're morally required to donate your resources as effectively as you can.[21] This might have surprising consequences. It might turn out, for example, that even if giving money to a less effective organization would be morally good, and even if it would have been permissible to do nothing, you're still doing something morally wrong by giving to the less effective organization.

I find this to be an interesting objection, but I am not moved by it. I deny the premise that you can bring yourself under a new obligation by deciding to do something morally optional. To put the point starkly: Nobody has a right to be saved by you.[22] If you don't have to go into the burning house to save anyone, then if— for some reason—you only save one person rather than two on your way out, you've done nothing morally wrong. We have the intuition that there is something amiss about this, but that is just because we find the psychology of such a person unintelligible. It makes no sense to leave someone behind, once you're in the house. However, saying something makes no sense is not the same as saying it's wrong. In the last section, I suggested that no one in the burning house has a right to your saving them. Entering the house does not give anyone any rights they didn't already have. So when you are inside, it remains true that no one has a right to your aid.

The upshot of all this is that even if you are an ineffective altruist, you do nothing morally wrong. Of course, it would be better to give effectively. And here I agree with the Effective Altruist: If you're in the business of doing something morally good, you might as well do as much of that as you can. But you don't run any moral risks by falling short.

So, are there any moral risks to Effective Altruism? My own view is that the central moral risk is in making unwarranted moral criticisms against others. For the most part, I'm happy to find that Effective Altruists present their arguments in terms of an opportunity to participate in the movement. And I think the movement of Effective Altruism does largely represent an extraordinary, collective moral project. At the same time, some philosophers encourage representing participation as a moral requirement.[23] And some economists have suggested that there is no moral risk with using the language of moral requirement, based on the observation that there is no apparent backlash to this language.[24] I think these ideas are morally risky.

Consider one final case, this one from the philosopher Hallie Liberto:

> Take John and Jane, young adults in college. Enthusiastic about John's upcoming football game, Jane promises John that she will have sex with him after the football game if his team wins, as an incentive for him to train harder and perform his best in the game.[25]

Should John accept this promise? Liberto says no. She thinks most us will agree that it would be wrong for any person to hold another person to a promise like this one. She calls this kind of promise "overextensive."

I agree. While Liberto thinks the explanation has to do with the content of sexual promises, I think it has to do with our absolute rights over our own physical person. Even though Jane has made a promise to John, he ought not hold her to it. She should retain the right to do as she chooses with her physical person in a way that is not constrained by the moral demands of others. Call this moral liberty.

Suppose you support Effective Altruism. Is it OK to demand other people join you, or to blame them if they don't? I believe it is not. We ought to respect the moral liberty of others. We should not suppose we have any entitlement to make demands on what separate persons do with their physical bodies. Insofar as Effective Altruists—or anyone else—are tempted to make such demands, they are in a risky moral business.

Question 12: Can Effective Altruism Be Part of a Meaningful Moral Life?

If there is no moral duty to be an Effective Altruist, that might sound like bad news for Effective Altruism. But there's good news, too. Decoupling Effective Altruism from moral requirement also helps to explain how participants enjoy a kind of moral security. The only real risk—as in many things—is trying to impose your values on others. As long as the Effective Altruist steers clear of that pitfall, then they merit our moral admiration.

If Effective Altruism is praiseworthy but not required, should we do it? Can Effective Altruism be part of a meaningful moral life?

Here my answer is that it depends on you. I am personally astonished by the Effective Altruist's capacity for sacrifice. For my own part, I do not participate in Effective Altruism. However, I really admire my colleagues (including the other writers in this volume) who do.

To put all this in a slogan, Effective Altruism is kind of like loving another particular person. If you do love someone, then sacrificing for them can make sense. But loving another person is morally optional.[26] It depends on how you feel.

I anticipate an objection to this view. Compare two kinds of personal projects. One person has a project of visiting every baseball stadium in the country. They invest considerable resources and time traveling from one city to another, checking ballparks off their list. In expectation, their project results in pleasant summer evenings and a warm sense of personal achievement. Another person has a project of contributing to eradicate malaria. They invest considerable resources in time in Effective Altruism. In expectation, their project results in saving the lives of strangers.

Once we hold the second project vividly in mind, doesn't it seem like it is much more important to become an Effective Altruist than to dawdle one's life away as the itinerant baseball fan? Surprisingly, perhaps, I will say no.

Consider a third case: Another person has a project of being a good parent to their children. They invest considerable resources and time in taking care of their family—time they could have spent either going to baseball games or maximizing their earning power to give to Effective Altruist causes. Here I think most people will admire the good parent just as they admire the Effective Altruist. These both seem like worthy causes. And yet, from the point of view of impartial value, the good parent might be deficient in more or less the same way the baseball fan is deficient. Granted, they might be promoting value more (even much more) than by going to games, but they might still do much, much better by contributing to the aid of strangers. Remember the drowning child case. Does it matter whether you ignore the drowning child because you don't want to ruin your shoes, or because you want to get home to read your own children a bedtime story? I see no difference.

The Effective Altruist might reply that one should be a good parent, but then dedicate remaining resources to Effective Altruist causes. This sounds plausible, but it is hard for me to see the principled case for it. Let's say that after discharging parenting responsibilities, one ought to be an Effective Altruist. If that were true, what would explain it? My guess is that it would be explained by the fact that one could promote overall well-being more by being an Effective Altruist than by using their resources in some other way. Or, if one prefers a contractualist formulation, it might be explained by the fact that the claim of the person one could save is morally more important than any other claim to one's resources. But if either of those explanations is true, it seems like they will undermine the idea that one ought to be the good parent *first*. After all, one could do better by utilitarian lights by diverting time and resources from one's children to those in much greater need. And by contractualist lights, the claims of strangers whose lives one could save seem greater and more urgent than the claims of one's children.[27] Here it is as an argument:

1. Either personal projects can be more important than saving strangers or not.
2. If personal projects are not more important than saving strangers, then spending leisure time taking care of one's children is not more important than saving strangers.
3. If personal projects can be more important than saving strangers, then being a baseball fan can be more important than saving strangers.
4. Spending leisure time taking care of one's children can be more important than saving strangers.
5. So, being a baseball fan can be more important than saving strangers.

Working backwards, I suppose most people will accept (4). We allow that if people love their children, then they can justify taking care of them even if it is not impartially value-promoting—indeed, even if it means not taking the actions needed to save an additional life.[28] (3) says if we grant that some projects we find worthy are OK to do instead of effective altruism, the reasons justifying them will also justify whatever we happen to care about—even the baseball game.[29] Once you let people do whatever they happen to care about, then the floodgates are open. (2) appeals to a similar thought. If you can't justify personal projects, then you can't justify extra care for your children. But since we can (given 4), and those are the only options (given 1), then it looks like we can justify any personal project.

What does all this mean for what's really important? My view is that importance is ultimately a psychological concept. What is important *for you* just depends on what *you* care about.[30] So if you care about Effective Altruism, then you are right to think that it is important. But if someone else doesn't care about it, then it doesn't make sense to say that they are wrong about what's important.

Long ago (in philosophical time), Thomas Nagel argued that there were two perspectives on reasons—an objective one and a subjective one.[31] I can think about what is good from the point of view of nowhere. From that perspective (roughly), there are only the morally relevant agents that exist and their welfare functions. But from my own personal point of view, I also have reasons to care about the particular person *who is me*, and the particular actions that are *mine*. Nagel famously doubted there was any reconciliation between these two standpoints.

I still feel the force of Nagel's idea. I am inclined to think that we should admire the Effective Altruist more than we admire the baseball fan (and maybe more than we admire the good parent). From the impersonal point of view, they are tracking the reasons in a way I am not. At the same time, I am persuaded that from the personal point of view, if the baseball fan says they care about baseball, then that settles the matter on what is important to them.

Can we make sense of these ideas together? I confess to being a little uneasy about the apparent mismatch between my considered judgment about how we should understand the idea of importance relative to a person, and my judgments about when moral praise and admiration are warranted. I can best make sense of the Effective Altruist as a kind of secularized analogue to Christian *agape*. Here I've followed a standard philosophical view in thinking that love justifies our partiality. If you love someone, you can do more for them, even if that means not doing as much for others. I've drawn out a slightly more controversial inference: You can love yourself, and so you can be partial to your own projects. You can also love people generally. Or, even if you don't, you can act as if you did (something Kant called *practical love*). R. Jay Wallace once observed that utilitarianism—despite being historically associated with atheism—has a kind of religiosity to it.[32] It asks a person to invest in a value in which they personally participate to only a very small degree. Of course, for the utilitarian that value is the aggregate welfare rather than the glory of God, but commitment to it still demands forsaking of one's own interests. And as religious folks have long understood, that is what devotion is all about.

Paul said that *agape* is the greatest of the gifts of God because it transcends all the others (1 Cor. 13). And we are now in position to say something kind of like that about altruistic commitment. If you can somehow care about all of humanity, then what is important to you personally will exactly match what is impersonally morally admirable or praiseworthy. There will be no tension between what you subjectively value, and what is objectively most valuable. Does that make for a meaningful life? It's up to you—and to me.[33]

Notes

1 For an application to the current question, see Sabato and Bar-Ilan 2023.
2 Weinstein and Ryan 2010.
3 Herman 2012; see also Macnamara 2013; Darwall 2006.
4 On the connection between feeling gratitude and intending repayment, see Sara B. Algoe and Jonathon Haidt 2019; McCullough et al. 2001.
5 For more detail, see Davis 2019.
6 It would be a morally permissible mistake. Harman 2016.
7 Some philosophers think you can be morally obligated to help, even if you don't have a directed duty to the specific recipient of your help. One strategy is to defend an imperfect obligation to help others, such

that there is a duty to help sometimes, if not in every particular case. Robert Noggle (2009) argues for an obligation to adopt a "propensity" to provide aid. Matthew Hasner (2014) holds that while there no requiring reasons to help, there are second-order requiring reasons to sometimes act on first-order justifying reasons to help. My view here agrees with Hasner that there are no requiring reasons to help, but differs about the second-order requirement.

8 Jessica Flanigan (2019) and Kieren Setiya (2022) have independently developed arguments supporting this premise.
9 Setiya 2021.
10 Grodeck and Schoenegger 2023.
11 Davis and Preece 2021; Karpowitz, Argyle, and Davis n.d.
12 Krishnarajan 2022; Landwehr and Harms 2020; Frederiksen 2022.
13 Caviola, Schubert, and Greene 2021.
14 Law, Campbell, and Gaesser 2022.
15 Deri, Stein, and Bohns 2019.
16 Reich 2013.
17 Gabriel 2015.
18 Nickel and Eikenberry 2009; Eikenberry and Mirabella 2018.
19 I don't share the moral objections to capitalism. I am sympathetic to Jason Brennan's (2014) view.
20 Cf. Horton 2017.
21 Cf. Muñoz 2021.
22 The most important paper elaborating this view is Thomson 1971.
23 Sachs 2019.
24 Grodeck and Schoenegger 2023.
25 Liberto 2017
26 Velleman 1999.
27 For the debate: Timmerman 2015; Logins 2016.
28 Cf. Setiya 2014.
29 Cf. Dorsey 2009.
30 Frankfurt 1988.
31 Nagel 1978.
32 Wallace 2004.
33 At least it makes for a life that isn't absurd (by Nagel's (1971) definition, anyway).

17
ARNOLD'S RESPONSE TO DAVIS

Question 10: Is There a Moral Obligation to Beneficence?

Davis contends that "there is no moral requirement to provide aid to others." Accordingly, "no matter how important or urgent it may be that others receive aid, and no matter how well positioned you might be to provide it, you don't morally have to do it."

Davis's piece is admirably clear and well argued. But his view has extremely implausible implications. Imagine that the entire state of Texas will be incinerated by hellfire unless you push a button marked "ABORT HELLFIRE." On Davis's view, while it would be very good of you to push it, you're not morally required to do so. This is because "choosing to provide aid to others—like all positive actions—is a matter of individual moral discretion."

Suppose you decline to push the button, thereby allowing some 30 million people to die. I submit that in failing to save millions at extremely trivial cost, you have violated a very serious moral requirement. Davis would disagree. For "what we do with our bodies is up to us," with the result that the person faced with needy strangers is "not guilty of any wrongdoing if [he or she] does nothing."

How does Davis arrive at such a counter-intuitive and, I think, incorrect result? We can reconstruct his core argument as follows:

1. You are morally required to F if and only if someone else has a right against you that you F.

2. Absent some special arrangement, nobody has a right against you that you provide aid to them.
3. Therefore, absent some special arrangement, you are not morally required to aid anyone.

I think we should reject both premises.

If premise (1) merely claimed that rights *can* generate obligations, I'd have no quarrel with it. Plainly, *one* way for me to be required to F is if you have a right against me that I F. But premise (1) goes further; it claims that the *only* way for me to be required to F is if someone else has a right against me that I F. That is, it claims that *nothing* but rights can generate obligations. And with that claim, I disagree.[1] A starving orphan on my doorstep may not have a right to the sandwich I was about to toss in the trash, and yet I think I have a moral obligation to give it to him anyway. Why? Well, because I regard as true something like Singer's famous principle that (in its more moderate formulation) *if you can prevent something very bad from happening at little cost to yourself, you are morally required to do so*.

Now, Davis would likely object here that there's a conceptual tie between *moral requirement* and *resentment*: you're only morally required to do what others could properly resent you for *failing* to do. But how, Davis would ask, could someone *resent* me for failing to F, if they don't have a *right* to my F-ing? My answer: the link between resentment and rights violation is not as tight as Davis supposes. It's quite possible to resent someone for failing to do what they should, even if nobody has a *right* to their doing it. Suppose you let my son die in front of you because you'd have to lift a finger to save him, and you don't feel like doing that. Further suppose my son has no *right* to your assistance. Even so, I assure you that I'll *resent* your failure to aid very deeply; and this resentment seems quite warranted to me.

Now, I've been assuming that the needy have no *right* to assistance from the well-resourced. This assumption aligns with premise (2)'s claim that, absent some special arrangement—like a contract or a promise or a special relationship—nobody has a right to be aided by you. But in fact this assumption, and hence Davis's premise (2), are eminently questionable. For there are such things as *human* rights, that is, "rights that individual persons possess simply in virtue of their humanity ... [and which] give rise to corresponding duties on others to ensure that the rights are protected and promoted."[2] Everyone has these rights, and everyone is—in some sense—responsible for securing them against standard threats.[3] One important human right, or so it is often claimed, is the right to subsistence, understood as the right to the material prerequisites for a minimally decent existence. If such a right exists, then Davis is simply wrong to assert that (absent some special arrangement) nobody has a right to your assistance. Alas, hundreds of millions and maybe billions of people have this right: namely, all those suffering from extreme poverty across the globe.

Would that I could continue: Davis's account is rich and thoughtful and I've barely scratched its surface. In particular, I haven't weighed in on his intriguing arguments from gratitude and noninterference. Rather than explore these, let me close by underscoring what I take to be my fundamental disagreement with Davis.

Pace Davis, not all moral requirements arise from rights. And even if they did, you'd still be morally required to aid strangers, because strangers have rights, human rights, to everyone's assistance (yours included).

But, bottom line, if this intricate argumentation leaves you a bit lost, simply return to my opening thought experiment. Davis says you're not morally required to lift a finger to save millions at no cost to yourself. I say you are. This is because, at root, if you can perform an "easy rescue," saving others without breaking much of a sweat, you are morally required to do so.

Which view seems more plausible to you?

Question 11: Is Effective Altruism Morally Risky?

In this section Davis advances a novel account of EA's "moral risks." This account has two main parts:

1. The first explains why EA isn't morally risky in ways other philosophers think.
2. The second argues that EA is morally risky insofar as it "makes unwarranted moral criticism against others."

In this brief reply, I explain why I disagree with both parts of Davis's account.

Let's start with claim (1). While other philosophers fret that EA undermines democracy and retards systemic change, Davis "[doesn't] share these worries." Effective Altruists, Davis explains, are engaged in *morally optional* activity. They're doing *more* than is morally required. How, then, could anyone demand *more* of them? They didn't have to help at all; but now that they are, it makes no sense to layer on new moral requirements (in particular, requirements to respect democracy and/or reform an allegedly unjust status quo).

Davis illustrates his argument with an example. Suppose you volunteer at your kid's school. That, notice, is supererogatory; nobody could demand such an activity from you. Now suppose, having volunteered, other parents impose various democratic demands on you, claiming that you must be transparent and accountable in your school activities. According to Davis, these further demands simply don't compute. "If you are already doing more than you morally have to," then "no one has any further claim on you."

I disagree. Supererogatory action *can* give rise to new moral requirements; or, perhaps better, an action can be supererogatory with respect to *some* moral requirements while running afoul of *other* moral requirements. I'm not required to bake you a cake; but supposing I do, is my cake-baking now blissfully free of moral constraint? May I comport myself, *qua* baker, just as I have a mind? Clearly not: suppose I bake you a cake filled with razor blades. That, surely, would be wrong.

So, baking a cake—a supererogatory action—*does* come with moral risk. It's awfully nice of me to bake it, but morally catastrophic if I kill you in the process.

A structurally similar moral risk looms over beneficent action, or so I argue in my second essay for this volume. There, I discuss "elite philanthropy," understood as big-dollar gifts given with various strings attached. For instance, Mark Zuckerberg donated upwards of $100 million to the Newark, NJ, school system, "on condition that the district embrace a slate of market-friendly reform proposals." This gift was, let's assume, surplus to moral requirements; Zuckerberg was not obligated to make it. According to Davis, this means Zuckerberg is immune to any democratic complaints. Not being obligated to give at all, how could he be obligated to give *democratically?*

That's like asking how a cake baker, not being obligated to bake at all, could be obligated to bake *safely*. I'm not required to bake you a cake; but *if* I so bake, I'd better do it without killing you. Similarly, Zuckerberg wasn't required to give $100 million to Newark; but *if* he so gives, he'd better do it without undermining democracy and/or political equality.

Simply put, supererogation doesn't eliminate moral risk. It's eminently possible to soar above the call of duty in one sense while plummeting below it in another. That, I'm afraid, is the situation of many elite philanthropists today.

I turn now to Davis's second main claim, namely, that the "central moral risk" associated with Effective Altruism is that of "making unwarranted moral criticisms of others." Davis is thinking here of Effective Altruists who "demand other people join [them]" and "blame them if they don't." Such behavior, Davis argues, amounts to "risky moral business" because it ignores the non-effective-altruists' "moral liberty," which entitles them—as, indeed, it entitles everyone—to "do as [they] please with [their] physical person in a way that is not constrained by the moral demands of others."

I would describe the risk landscape differently. To see this, consider another example.

Assume, *arguendo*, that (as Davis asserts) people are not morally required to aid anyone else. So, you're not required to, say, save my son from drowning in a shallow pond, even if you could save him at basically no cost to yourself. Now suppose I *demand* that you save him. (Maybe I'm on the phone with you, and you're the only one who can reach him.) "He's right there!" I plead. "Go get him!" In response, silence. So I up the pressure: "Save him! It's a moral requirement!" Imagine you decline, politely citing your moral liberty, and then hang up. My son drowns.

On Davis's analysis, *I'm* the villain in this story; *I'm* the one incurring all the moral risk. Yes, you let my son drown for no good reason. But that's not morally wrong, since (on Davis's view) he had no right to your assistance and you are not required to aid anyone unless they have a right to that aid. I, on the other hand, had the temerity to *issue an unwarranted moral demand*: and that's "risky moral business," insofar as it intrudes upon your moral liberty.

I must part ways with Davis here. The human condition is freighted with moral risk. Some, I grant, involve making excessive demands on others. But such risks seem rather marginal compared to, say, the risk of failing to perform so-called "easy rescues"—failing to prevent extremely bad things from happening, at trivial cost to oneself.

Like Davis, I say: "Don't be a hectoring moral puritan." But unlike Davis, I add: "*Also*, and *more importantly*, don't be the sort of person who lets a kid drown in front of you for no reason."

Question 12: Can Effective Altruism Be Part of a Meaningful Moral Life?

You're not *required* to be an Effective Altruist, according to Davis. But that leaves open whether you *should* be EA. Effectively benefitting others might be morally important or praiseworthy even if it's not morally required. Indeed, doesn't EA seem much *more* important and praiseworthy than many other pursuits? It's fine to tour baseball parks, but much better—surely—to cure malaria (to use Davis's example).

Davis says: not so fast. Following Nagel, Davis identifies two perspectives. The first is the objective, impersonal "view from nowhere." The latter is the subjective, personal view from behind your eyes. If I read him correctly, Davis concedes that, judged impersonally, EA is indeed more important and praiseworthy than touring baseball parks (or any other parochial, personal project you care to name). But he resists declaring EA more important in some overall sense, because, from the personal perspective, "what is important for *you* just depends on what *you* care about."

The upshot? For Davis, you always have an objective, impersonal reason to be EA: namely, being EA is an excellent way to benefit others. But whether you have a personal, subjective reason to be EA depends on whether you happen to prioritize beneficence over more self-interested pursuits. If so, great: "there will be no tension between what you subjectively value, and what is objectively most valuable." If not, then you're in a tough spot, subject to contradictory imperatives from competing moral perspectives. ("I know I *should* donate to the anti-malaria campaign. But what I really *want* to do is see Fenway Park.")

This all seems correct to me. But Davis gives the impression that this conflict between perspectives simply is what it is; about it, nothing can or should be done.

If I were telling the tale, our story wouldn't end with irreconcilable conflict between normative perspectives. Instead, I'd explore strategies for intentionally cultivating overlap between what one finds personally meaningful and what's impartially best. And I'd ask whether people are obligated to pursue these strategies. Are you morally required to change your outlook, so that acting beneficently becomes more personally meaningful?

It seems to me that one can deliberately engineer psychological changes, such that one comes to *want* to act beneficently. One then finds deep meaning in doing what's objectively good.

Any pet owner or parent knows this possibility well. Caring for a vulnerable creature requires enormous amounts of self-sacrifice. But we're wired for it. Lavishing attention and care upon a helpless little baby or puppy or whatever isn't *fun*, but subjectively, it's among the most meaningful things most people do in their lives. Hence, a happy overlap between Nagel's spheres obtains in this case.

What I'm suggesting is that this area of overlap can be deliberately expanded. Through contemplative practices like "metta" (loving-kindness) meditation or prayer, one can come to care about—and even love—more and more beings. Serving their good thus becomes serving your own.

Imagine Davis's baseball fan gets really into prayer, or loving-kindness meditation, or whatever (I don't mean for these examples to be exhaustive). "Cure malaria, or visit Fenway?" we then ask him. He's transformed his psychology such that the former strikes him as the thing to do *subjectively* as well as objectively. It's what reason supports from *both* of Nagel's perspectives.

At this point, I can almost feel the audience rolling its eyes: "Another professor advocating a pie-in-the-sky, hippie solution to the world's problems." To be clear, I'm not claiming that many people *will* embark on this path of empathetic broadening. My point, rather, is twofold: first, such a path undeniably exists; second, if people take it, they can reduce the conflict between what's subjectively meaningful and what's impartially best.

This raises an interesting question: are people morally obligated to take this path, obligated to develop Davis's *agape* for an ever-widening circle of beings? Perhaps the "woo" factor of metta and prayer distracts from the core issue here. So let's boil this down to essentials. Davis's baseball fan finds meaning in personal projects with little beneficent upshot. But suppose he could take a pill to shift his preferences in a beneficent direction. Then he'd find great meaning from doing good. He'd gladly choose curing malaria over touring ballparks.

Should he take the pill? I'd be interested to hear Davis's thoughts.

Notes

1 I draw here from Larry Temkin, "Thinking about the Needy, Justice, and International Organizations," *The Journal of Ethics* 2004, 8, 4, 349–395.
2 Jones 2013.
3 See Shue 1996.

18
BRENNAN'S RESPONSE TO DAVIS

Question 10: Is There a Moral Obligation to Beneficence?

Psychologists claim that our judgments are biased by an anchoring effect. In famous experiments, psychologists ask subjects to estimate how much a person's height deviates from an arbitrary "benchmark." The choice of benchmark changes their estimation. Subjects guess a high number when asked how much the target is shorter than six feet and guess a low number when asked how much the target is taller than five feet.[1] They guess different numbers when not given benchmarks.

Philosophy has benchmark biases too. When theorizing about beneficence, I see three benchmarks which philosophers often use. Some philosophers, such as Thomas Hobbes, Gregory Kavka, Jean Hampton, David Gauthier, or David Schmidtz, start from the benchmark of self-interested means-end reasoning, and then try to build a case for altruism from this starting point.[2] Most end up concluding there are duties to help others, but the benchmark tends to induce them to conclude such duties of beneficence are limited.

Others, such as Peter Singer or Peter Unger, start with the presumption of equal and impartial concern for all lives and everyone's welfare and presume that welfare matters most.[3] They then try to carve out, or at least consider attempts to carve out, space for personal prerogative and close personal relationships. They end up concluding we have quite limited space for personal prerogative and quite extensive obligations of beneficence.

In contrast, some philosophers (e.g., W. D. Ross, Michael Huemer, and I) tend to treat intuitive commonsense ethics, the ethics we actually live by, as a starting point. Commonsense holds that we have imperfect duties of beneficence, with wide latitude and freedom in how we satisfy those duties. Further, common sense holds that the easier it is to exercise beneficence (say, by being richer), the stronger the duties become. While we sometimes end up rejecting parts of commonsense ethics (e.g., Huemer and I both reject nationalism and endorse open borders), in the case of beneficence, we haven't budged too much from the benchmark of common sense.) Still, if common sense was the wrong starting point, then it's not interesting whether, as we think, there aren't strong grounds to budge from that point.

It sure would be informative and fortuitous if all three starting points or benchmarks led to the same conclusions about beneficence. But they don't seem to. And it's unclear which starting point is correct. Each starting point could be seen as question-begging rather than neutral.

Davis does not take for granted that the commonsense view—that we have imperfect duties of beneficence toward others—is correct. He tries to defeat this commonsense view by noting *other* features of commonsense ethics which seem incompatible with it. First, he notes that we feel gratitude—indeed, think we *should* feel gratitude—when people save drowning children. If we feel gratitude toward someone, he thinks this suggests the actor went above and beyond the call of duty, rather than doing their moral minimum. After all, I don't feel gratitude toward Davis that he never mugged me. Davis also notes that we cannot *force* someone to save a drowning person. He thinks this suggests that there is no obligation to save them.

Regarding gratitude, Davis is correct about people's reactive attitudes. Suppose I pull a drowning child from a shallow pool, incurring no risk of death or physical harm to myself. Let's say it takes three minutes of my time, but ruins my $500 blue suede shoes. People would not react by saying, "Ah, Brennan did nothing more than the moral minimum. He did the least anyone should do. Praising him for *saving* that child is no more warranted than praising him for not eating the child." Rather, they would treat me like a hero. The child's parents would feel grateful to me forever.

This is some evidence that such behavior is not really obligatory, despite people *saying* it is. However, it does not seem decisive. *Pace* Davis, I am unsure that the reaction of gratitude indicates an action is supererogatory rather than obligatory.

One possibility is that we have developed norms of expressing gratitude toward people for performing certain actions—even when those actions are obligatory—if we know most people shirk on those duties, if those duties are demanding, or if when gratitude provides useful encouragement. For instance, many people claim there is a duty to vote, but still hand out "I voted!" stickers and praise people for voting. We thank veterans for their service even when we think serving was obligatory. We even thank veterans who didn't even choose to serve but were instead drafted against their will. The goal might be to induce more of this behavior, even though the behavior is required. We get more of what we reward, after all.

More importantly, there seem to be many common cases where we thank people for performing their duties. We thank doctors, nurses, and teachers for their care, even though caring for us is their job, which they are contracted and paid to do (sometimes handsomely). We thank our spouses for birthday gifts, though such actions are expected. We express gratitude towards our parents and grandparents for their care, even when such care is required from them. So, again, I am unsure whether the presence of gratitude indicates an action is supererogatory rather than obligatory.

Davis's second argument is that, except in special cases, we cannot interfere with others to make them act beneficently. He thinks this indicates that these actions are not obligatory. I find this objection less troubling than the first.

For one, I think that most of our duties are not enforceable (or at least not enforceable by most other people). For instance, suppose Mark is about to cheat on his spouse. Even though he has a clear moral obligation not to do so, I may not use violence to stop him. Further, unless his spouse is my friend or close to me, it's unclear that I may scold him for cheating or inform his spouse. He should not cheat, but I lack standing to do much about it.

Or, suppose there is an obligation to honor Jesus. It doesn't follow that I am permitted to browbeat people into worshipping him, or use the state to force them to worship him. Or, suppose there is an obligation to improve one's talents and virtue. It doesn't follow I or anyone else can push you around or force you to do so.

We can think of hundreds of examples like this. Many of our moral obligations are not enforceable by most other people. Some might not be enforceable by anyone.

Davis says, "To say that an action is required is to say that it is someone else's business." I disagree. Morality is pervasive, but morality also constrains how much moralizing we may do to each other. We are obligated to avoid being moral busybodies who constantly give each other moral advice, point out each other's moral wrongs, and push each other around.[4] To borrow an example from Justin Tosi and Brandon Warmke, suppose at grocery store, that a child pesters their parent to buy some treat. Finally, the mother snaps at the kid. Her face flushes red with embarrassment. The mother's behavior is wrong, but it's not your place or mine to chastise her, offer parenting advice, or remind her of her duties.

Question 11: Is Effective Altruism Morally Risky?

Davis has some great responses to common objections to Effective Altruism. I think these responses are sound and the objections fail.

However, despite being a proponent of Effective Altruism, I accept that EA is morally risky in other ways.

One way it is risky is that people might misapply the standards. This does not make the standards *incorrect*; after all, if you misapply some standards, then by hypothesis the problem was not that the standards but your misapplication. By analogy, consider the issue of defensive violence. Politicians who start unjust wars frequently claim (sincerely) that they were acting in justified self-defense, though they were not. The problem is not that the standards of just war theory are wrong, but that flawed people are bad at applying the standards to real-life cases.

Still, even if this response (it's the people, not the standards) saves EA as a theoretical project, it makes trouble for it as a practical project. EA proponents hope to recruit individuals into their camp. They recognize that most people who wish to adopt EA ideas don't have the time or wherewithal to think through the complications and applications at great length themselves. They cannot all be junior philosophers, development economists, charity evaluators, and the like. So, they need to rely on others' ideas and testimony.

To that end, various EA theorists have developed resources to help others apply EA principles. For instance, 80000hours.org provides advice on how to increase one's positive impact through one's career. GiveWell.org recommends high bang-for-the-buck charities. GivingWhatWeCan.org provides resources to reduce the cognitive and motivational strains of giving.

I might have some quibbles with some of their particular recommendations, but my worry about risk is broader. Even if those organizations are perfect right now and give the best possible advice, organizations have a life of their own. Over time, any given charity, including charities whose goal is to assess other charities, tends to be captured by careerists who care more about their own advancement than the goals of the group. For instance, two of us authors were involved with one organization (which I won't name) that has started doing less and less with an ever-growing budget. The current leader seems to view the job as a stepping stone to a university presidency. In general, the people who found charities might be true believers, but after a decade, half their staff are moderates or people who just needed a job. Charities suffer from mission creep—the tendency to expand in ever more marginal activities as a way of justifying their budget. Finally, even good people make mistakes. 80000hours.org might be giving excellent advice now, but could give terrible advice next year, thanks to cognitive errors.

To be clear, this is nothing special to EA. EA organizations are risky because they are *organizations*, not because they are *EA* organizations.

Another risk of EA is that it could become a substitute for organized religion, one that—as religion sometimes do—swallows the effective altruist's life and renders them miserable. As Sigal Samuel says:

> [EA] actually mirrors religion in many ways: functionally (it brings together a community built around a shared vision of moral life), structurally (it's got a hierarchy of prophet-leaders, canonical texts, holidays, and rituals), and aesthetically (it promotes tithing and favors asceticism). Most importantly for our purposes, it offers an eschatology ... in the form of its most controversial idea, longtermism ...[5]

It wouldn't be surprising if among certain groups, EA becomes a cult in which people bully each other to prove their purity, demand as much sacrifice as possible for the cause, abuse their power and abuse others' good intentions, and render members miserable.

Indeed, we've seen some examples of this behavior: Sam Bankman-Fried defrauded people out of billions and duped prominent EA leaders. Think of how much money was lost because EA proponents were wrong about him.

Again, none of this is special to EA. These concerns apply to all social, intellectual, religious, or political movements. People make empirical or philosophical mistakes. Movements get captured by fools, power- and status-seekers, and frauds. They overreach.

By analogy, consider how Ayn Rand's philosophy demands that people think for themselves. Yet, many biographies indicate her close followers were conformists. Or consider that communists are supposed to be philosophical egalitarians, but in practice are hierarchical totalitarians.

EA opposes and seeks to eradicate the problems that make non-profits and civil society go bad, but that doesn't mean it won't itself go bad. Organizational problems run deep; it's not simply that most organizations lack good philosophy or fail to write up cost-benefit analyses.

Question 12: Can Effective Altruism Be Part of a Meaningful Moral Life?

What makes life meaningful and what makes it moral are not the same. Paul Gauguin, Jean-Jacques Rosseau, and Mao Zedong led meaningful lives—not just *significant*, but meaningful—but these were morally bad lives.

Conversely, leading a morally good life is not sufficient for it to be particularly meaningful. Imagine some freedom fighter tries to assassinate Mao at the height of the Cultural Revolution. The plot fails, the person is tortured and executed, and China continues its totalitarian oppression. That freedom fighter is admirable, but their utter failure makes me question whether their life is meaningful.

There all sorts of projects worth spending a life on. For instance, Davis, Chappell, and I each agree EA is a good thing, despite our disputes. It seems unlikely that any of us would have chosen to become academic philosophers were we trying to maximize our positive impact. But each of us thinks philosophy helps to make our lives meaningful and good.

Still, this all depends on what it means for a life to be meaningful. That's its own can of worms. People claim all sorts of things make their life meaningful. Some say long-term projects, such as writing a book, making art, discovering new scientific knowledge, landing a lead role on Broadway, and whatnot, make their life meaningful. Some think meaning comes from communion with God. Some think it is about building a loving family life. Some might think meaning comes from serving others as best they can, say, by joining Doctors without Borders or serving in the military in a just war.

For any of these activities or projects, we can find people who have achieved them and yet fail to find them meaningful. As Thomas Nagel noted, we "always have available a point of view outside the particular form of our lives, from which the seriousness [by which we take things] appears gratuitous."[6] He says the two points of view—the internal point of view from which things seem important and the external point of view from which our striving seems pointless—clash and make our lives seem absurd. The good news is that if our lives are fundamentally absurd (and I'm not saying they are), then the absurdity of them is nothing to take too seriously either.

A meaningful life need not be happy. To take a fictional case, Jean Valjean's life was highly meaningful, but it was mostly miserable.

Let me recast Davis's question. Can EA be part of a life worth living, that is, a life we would want to experience ourselves or which we would wish upon on our children?

To elaborate on what I mean by that, consider that some people have a view of morality whereby being a moral agent, an agent who does the right thing, more or less guarantees misery. On their view, morality is so demanding that to act rightly means living in perpetual service of others. We must respond to one tragedy after another. Love, hobbies, friendship, knowledge, the development of one's skills and personality take a backseat to service, and are permitted only when they are useful to ensure we serve others better. On this view of morality, to be a moral agent is like a curse.

Jean Hampton wrote a famous critique of pathological altruism. She describes Terry, a pregnant woman who burdens herself to the point that her exhaustion killed her unborn children and rendered her an ineffective parent to her older children. Hampton says that while others bear some blame for Terry's situation, Terry does as well. Terry failed to treat herself with proper self-respect. She failed to recognize her own value.[7]

Note that Terry is not a counterexample to utilitarian altruism. After all, Terry was an ineffective altruist. She was all cost and no benefit! Even a utilitarian would say that we must take good enough care of ourselves—so that we can maximize the good we do for all. Just as overgrazing a field can reduce its overall output, so can overworking a person. Remember, the utilitarian adds, that the good of all includes our own good... as one of the eight billion people whose lives and welfare matter equally.

Hampton isn't trying to peg Terry as a counterexample to Peter Singer. Her point is that Terry mistreats herself, and what makes her self-treatment bad is not simply that her self-sacrifice fails cost-benefit analysis.

Hampton thinks it's a mistake to presume that all altruistic actions (even effective ones) are morally good or that selfish actions are morally suspect. She argues that "we must define a new conception of morality which recognizes that any 'altruistic' behavior is morally wrong when it prevents one from paying moral respect to oneself."[8]

Consider Ursula Le Guin's short story, "The Ones Who Walk Away from Omelas." The story describes an idyllic, almost utopian society. There is no war or disease. Everyone is healthy, beautiful, and happy. However, Omelas has a secret. A single child is imprisoned in a closet, filthy, starved, tortured, and afraid. Through dark magic, torturing the child is what makes the city so splendid. At some point in their education, each citizen of Omelas is brought to see the child. Le Guin ends by saying that every night, a few citizens walk away from Omelas.

They walk away because Omelas, despite maximizing utility, is an evil and unjust city. It is wrong to exploit the child like that to benefit the rest, though the rest benefit far more than the child suffers.

Now, the child is not there by choice. She did not agree to sacrifice herself so that others may flourish. The citizens of Omelas treat her like a tool. What if instead Omelas offered you the choice to place yourself in the closet, living in filth and pain so that others may flourish? Would you be *obligated* to consent? Would it be heroic and good?

I suspect Hampton would say no. What's makes it wrong to treat the child that way is not merely that she didn't consent. A person who makes themselves an object of exploitation mistreats herself.

Hampton says you should regard yourself (and others!) as having a kind of intrinsic worth that resists seeing yourself as just one mouth among many, whose welfare is interchangeable with any others. It means resisting seeing yourself as an object to be exploited for the benefit of others, such as by undertaking an almost slave-like existence of continuously working to save drowning children, taking breaks only to ensure one maximizes one's overall output. Further, she thinks we not only have a right but a self-regarding obligation to develop ourselves into distinct and particular people. We have a standing permission to choose ourselves over others, and not only when our needs are more pressing than theirs.

Notes

1 Tversky and Kahneman 1974.
2 Hobbes 1651; Kavka 2021; Hampton 1987; Gauthier 1987; Schmidtz 1995.

3 Singer 1972; Unger 1996; Tan 2004.
4 See Tosi and Warmke 2024.
5 https://www.vox.com/the-highlight/23779413/silicon-valleys-ai-religion-transhumanism-longtermism-ea
6 Nagel 1971, 719.
7 Hampton 1993.
8 Ibid., 146.

19
CHAPPELL'S RESPONSE TO DAVIS

Question 10: Is There a Moral Obligation to Beneficence?

Davis offers two arguments—from gratitude, and from noninterference—for the striking conclusion that we have no positive duties to aid others. I think both arguments are questionable. But my primary interest is in questioning the significance of the conclusion.

Gratitude

The argument from gratitude assumes that aid warrants gratitude, whereas doing one's duty doesn't, so aid cannot be one's duty. As Davis anticipates, we may question the assumption that doing one's duty never warrants gratitude. Although he notes that "you don't have to feel grateful to someone who merely *withholds* from violating your rights," not every duty is so trivial. If positive action is required to respect your rights, I think that can warrant gratitude. (The fact that we needn't feel grateful to a terrible person who must exert great effort to refrain from abusing us doesn't show that fulfilling difficult duties *never* warrants gratitude. It might matter whether the duty would be difficult even to a decent person.)

For example, I think we should generally feel grateful to our (non-abusive) parents, even assuming they did no more than their parental duties—for that alone is no easy task! Likewise for academic supervisors and others who are duty-bound to assist us in sometimes quite onerous ways. In general, I think we should feel grateful when others help us. We can appreciate the effort that went into satisfying an onerous duty, and be grateful that they were willing to do this for us. It shows an admirable degree of good will on their part. It may also be true that it would have been wrong for them to do otherwise, but of course a grateful person will not focus so much on this negative side of things.

But perhaps there's more to it. Suppose your dutiful advisor clearly didn't care about you, and rather resented every moment they had to spend advising you. They would no longer merit your gratitude. Likewise if someone has to be harangued into saving you from drowning in a shallow pond. Gratitude is called for when we are helped *from benevolent motives*. Merely (reluctantly) doing the right thing is not enough. So aid doesn't necessarily warrant gratitude. Both key assumptions of Davis's argument turn out to be mistaken.

Noninterference

The argument from noninterference assumes that we may permissibly force people to do their duties, but we may not force people to aid, so they cannot have a duty to aid. Again, I think both premises are questionable.

Whether you can force people to do their duty depends on the details of both the duty and the proposed interference. There are good general reasons not to interfere with others' lives, especially via overt threats of violence. For one thing, they will often know relevant details that outside observers don't. And it would undermine social trust and co-operation to be constantly meddling in each other's lives. We may judge parents who spank their children to be acting wrongly, without thinking it's our place to interfere. For another example, it's surely wrong to cheat on your romantic partner—they could reasonably resent you for it—but it hardly follows that they could permissibly threaten you at gunpoint to prevent it!

So reasons for blame or resentment don't automatically translate into reasons to interfere via threats of violence. But I also think we *may* force people to aid by *some* means. Davis invites us to imagine a capable swimmer who refuses to save a drowning child, but *would* save them if forced into the water. We may not pull a gun on them. But if they're standing on a trap door on the dock, I think you may (and should!) pull the lever to dump them into the water. They have no grounds to complain about this minor indignity when a child's life is on the line.

I also think that, if given the option, we may (and should) vote for higher taxes that would go toward effective global aid. This may be so *even if* we're individually permitted to not donate to charity.[1] (Similarly for anti-factory-farming animal welfare regulations compared to individual meat-eating.) This is because voting for good outcomes is much easier, psychologically, than initiating the relevant changes in one's personal life. So there's less excuse for not supporting impartially better outcomes at the policy level. Political selfishness seems an especially egregious form of selfishness, whereas I think we're all a bit akratic in our everyday lives.

Significance

Suppose Davis is right that we have no positive duties of aid, in the sense of wrongdoing that's tied to warranted resentment. It may yet be the case that we have most reason to help others. As Davis agrees, "It does seem obvious that we ought to save the drowning child, that this is *the thing to do*." Practical deliberation that concludes in beneficent action is *correct*, whereas selfish acts are importantly *mistaken*—contrary to reason—whether or not it's a mistake of a sort that others could reasonably blame us for.

As social creatures, we may be especially concerned to not fall so low that others have grounds to resent us (or threaten our social standing through public censure). But as rational agents, we should also try to be guided by good reasons, and make good decisions. The fundamental question of ethics is not *what can I get away with?* but *what should I do?* Or, in other words, *which option is most worth pursuing?*

So while I'm inclined to think that the global poor *could* reasonably resent us for drinking champagne as we watch their children drown, I don't think the case for beneficence depends on this assumption. Perhaps we could get away with doing nothing. But whether or not we're *required* to act beneficently, it is *the thing to do*.

Question 11: Is Effective Altruism Morally Risky?

Davis generously defends Effective Altruism against traditional charges of "moral risk," but suggests there is a risk of placing mistaken demands upon others. I think there is much more risk all around, but I'm less concerned about the one Davis highlights.

Risky Demands

I'm sympathetic to the suggestion that we should generally respect others' moral liberty, and "not suppose we have any entitlement to make demands" on them. But there is also a moral risk to complicity in moral complacency. If the choice is between being uncomfortably confrontational or continuing to allow millions of children to die each year from easily preventable causes (and billions of animals to be tortured on factory farms, and the entire future of humanity to be threatened by existential risks), then a little infringement on the moral liberty of the most privileged people who have ever existed sure strikes me as the lesser evil.

I don't know whether that is the choice we face. I hope that framing Effective Altruism as an opportunity rather than an obligation will tend to prove more motivating: you catch more flies with honey, and all that. But I do think it would be a bit... precious... for the global rich to complain about the *imposition on their moral liberty* of others demanding that they spare a thought (and a modest portion of their income) for the lives they could easily save while still living far more comfortably than almost anyone else who has ever existed.

To bring out this point, just imagine seeing someone calmly picnic while two toddlers drown in shallow ponds behind him. "Quick, behind you!" you call out, as you rush to the second pond. "Oh, yes, quite disrupting the ambience," the man sniffs disdainfully, and turns away, before adding darkly, "I trust they'll quiet soon enough." Now what is the greater risk: that you mistakenly demand that he save the child, or that you mistakenly acquiesce to his apathy and allow a child to die unnecessarily?

Even if there is some risk of demanding too much beneficence from the global rich, it strikes me as a comparatively minor moral harm, and not one that should fill our moral attention. (Compare worries about anti-white racism, or about immigrants committing crimes: it's true enough that those are bad things that sometimes happen. But it would be a mistake to allow these comparatively minor problems to distract us from addressing much graver social injustices.)

So if—implausibly—haranguing apathetic privileged people was an effective way to get them to do more to help those in need, I don't see that they would have any grounds to complain. And we should care much less about their petty complaints than about those who would otherwise suffer and die needlessly.

Uncertain Choices and Opportunity Costs

There's a different kind of risk I'm more concerned with, which is just that cause prioritization raises a lot of high-stakes empirical questions that are very difficult to answer. Effective Altruism is not *distinctively* risky here—reflecting on the questions presumably increases our chances of getting the answers right, compared to sticking our heads in the sand—but it does at least increase the *salience* of the risks, in a way that can be psychologically distressing.

For example, some scientists have proposed insect farming as an alternative protein source to replace factory-farmed animals (at least on the margin). Depending on whether the insects are conscious and suffer, such a transition could be either very good or very bad (due to the staggeringly large numbers involved). Either option faces the risk of possibly causing vastly more suffering than the alternative.

Or consider the tradeoffs between aid to the global poor vs investment in AI safety. There's a small chance that the latter will prevent a global catastrophe, while the former saves lives with near-certainty. There's a (potentially) huge moral cost to underfunding *either* of these two moral goods. Some on each side suggest that their preferred cause is many orders of magnitude more important than the other, making misallocation morally disastrous. But it's far from obvious which side is right, or what the ideal allocation would be. Some kind of balance between the two seems a reasonable response to our first-order uncertainty, but even this is uncertain.[2]

In such circumstances, I think it makes most sense to adopt a liberal attitude of appreciating all reasonable, good-faith efforts in this vicinity. It's unreasonable to demand perfection, especially when it's so unclear what the best option would even *be*. But I do think it is generally good to at least *try*, and simply accept that even our best efforts are fallible.

Importantly, it's not as though *not even trying* lowers the moral risk here. Quite the opposite: the risk of *failing to choose the best option* (when the moral stakes are high) is surely vastly higher when not even *trying* to identify and select the best option.

The Risk of Moral Fanaticism

Finally, one might worry that Effective Altruism could inspire moral fanaticism. For example, while the details remain in dispute, some have speculated that FTX executives (who prominently self-identified as Effective Altruists) convicted of fraud may have been partly motivated by their desire to redirect ever-larger amounts of money to charitable ends.[3]

It would be awfully ironic if it turned out that moral motivation was, on the whole, a force for evil in the world, simply because of how extraordinarily bad the rare fanatic ends up being. I'd be surprised if the bad really does outweigh the good, but I have no special expertise in addressing such empirical questions. What I can say is that these sorts of examples help to illustrate just how important it is *in practice* for our moral efforts to be constrained by respect for others' rights—even on purely utilitarian grounds. And this is a familiar observation in the utilitarian tradition. For example, in his classic 1984 defense of a utilitarian basis for rights, Gibbard writes that "Pure Machiavellian benevolence has poor utilitarian backing as a standard for a person of anything close to human possibilities and limitations."[4] Or we can go all the way back to J.S. Mill: "People talk as if... at the moment when some man feels tempted to meddle with the property or life of another, he had to begin considering for the first time whether murder and theft are injurious to human happiness."[5]

Machiavellian altruists are cause for concern. As are other Machiavellian individuals. But the problem there is the Machiavellianism, not the altruism.

Question 12: Can Effective Altruism Be Part of a Meaningful Moral Life?

Our personal projects and deep concerns help to make our lives feel meaningful. But they may fail to track our intellectual judgments about what is most *objectively* important. Mismatches between objective evaluation and subjective commitment raise interesting questions. For example:

(1) How should we prioritize between objective and subjective importance?

Davis suggests that the subjective perspective trumps: "if someone else doesn't care about [saving lives], then it doesn't make sense to say that they are wrong about what's important."

That seems too strong to me. I think it depends on the details. Common sense grants us wide-ranging, but not absolute, personal prerogatives. It can seem reasonable to prioritize being a good parent over saving a small number of strangers (when you cannot do both). But if you sufficiently lower the value of the personal project and/or raise the impersonal stakes, I think we will eventually find that our optionality is limited. Someone who cares more about the scratching of their finger than the destruction of the whole world is simply unreasonable,[6] and truly *mistaken* about the comparative importance of these two ends.

The idea that desires, and not just beliefs, can be mistaken or unreasonable may seem a surprising one. But consider Parfit's *Future-Tuesday Indifferent* agent, who "would choose a painful operation on the following Tuesday rather than a much less painful operation on the following Wednesday."[7] The imagined agent knows he will subsequently regret it, but simply doesn't care—about either his future agony or the associated regret. Such an agent seems less than perfectly rational. Many of us would probably describe such a pattern of concern as "senseless" or even "crazy." As Parfit sums up his case: "Preferring the *worse* of two pains, for *no reason*, is irrational." Less extreme cases of misguided priorities may also be less than perfectly rational.

We (plausibly) have some reason to care especially about our own children, and many other of our personal projects—even baseball. But we have *more* reason to care about our children than to care about baseball. So it isn't true that any time we could prioritize our children over an impersonal good, we could just as reasonably prioritize baseball over it. Not all personal projects are equal: some yield stronger reasons than others.

If we could *always* reasonably prioritize our subjective concerns over what's objectively important, that would provide an easy answer to our first question. But I do not think it has such an easy answer.

We may make further progress on a distinct (yet related) question:

(2) In selecting personal projects, do we have more reason to *develop* concern for what's objectively important?

Our first question took for granted that our subjective concerns were *already settled*. But this is not always the case. Sometimes we develop new interests, and new concerns. And this is not just something that happens *to* us, like a robot passively awaiting programming. We can reflect on our values, and on how we live our lives, and act in ways that bring the two into closer alignment. We can influence our (future) subjective commitments through our choices of what to attend to, who we engage with, and what we sign up for.

You could develop a new hobby by signing up for a class, reading books on the subject, and making friends with others who already have the interest in question. You could probably develop an interest in Effective Altruism by similar means: reading *Doing Good Better*, engaging on the EA Forum, starting or joining an EA student group at your university, or signing up for a trial pledge with *Giving What We Can*.

If you're not antecedently committed to either, which do you have more reason to develop an interest in: baseball, or Effective

Altruism? The latter is obviously favored by the objective perspective, and there's not yet any subjective reason against it. If you commit to baseball, that will *generate* a conflict between the objective and subjective perspectives. But, as we're imagining things, there's no conflict yet. So why create one?

There seems something normatively ideal—a kind of deep integrity—to aligning one's subjective perspective with what (one recognizes) objectively matters.[8] Given the human condition, this goal is not perfectly attainable: we will inevitably find ourselves with self-centered concerns that aren't impartially justifiable. But we can at least take steps to mitigate the size of the gulf between our values and our lived reality. Giving *some* non-trivial weight to effectively helping others in practice seems a crucial means to advancing our moral integrity.

A popular alternative would be to self-deceive yourself about what objectively matters. You could bury your head in the sand, refusing to recognize all the preventable suffering in the world. You could deny that foreigners, or non-human animals, or future generations matter. Plenty of people make such claims; you'd be in good company—or company, at any rate.

But is that really the sort of person you want to be? Isn't there something to be said for confronting reality head-on, without comforting illusions? Even though there are limits to how much you're willing to do to address the world's problems (as there are for us all), couldn't you justifiably feel that your life was *more* meaningful if you were seriously contributing to making the world a better place? It needn't be an all-consuming project: you can likely find room in your life for baseball too, if you want. But if you do both, it isn't hard to predict which one you'll ultimately find to be most meaningful.

Notes

1 This is orthogonal to Davis's principle. He assumes that if X is S's duty, then we may force S to do X. Some are drawn to the related, but distinct, principle that if X is *not* S's duty then we *may not* force S to do X. I'm now suggesting that we should also reject this distinct principle. I think there's no essential connection, in either direction, between duties and warranted force. (They may correlate, but only imperfectly.)

2 For an overview of the scholarly debate over how to respond to moral uncertainty, see MacAskill, Bykvist, and Ord 2020.
3 Others suggest that the fraud may have stemmed more from incompetent book-keeping than deliberate strategy. See, e.g., Lewis 2023.
4 Gibbard 1984.
5 Mill 1863, Chapter 2. Available at: https://www.utilitarianism.net/books/utilitarianism-john-stuart-mill/2/. For more on this theme, see: https://www.utilitarianism.net/utilitarianism-and-practical-ethics/#respecting-commonsense-moral-norms
6 *Contra* Hume 1739, sec. 2.3.3.6.
7 Parfit 1987, 124.
8 See Railton 1984, sec. ix on the costs of alienation from morality itself.

20
DAVIS'S RESPONSE TO ARNOLD, BRENNAN, AND CHAPPELL

I have gotten a lot from the responses to my essay. Each one is clear, full of philosophical content, and fun to read. Let's start with the biggest issue. My co-authors think it's crazy that I believe morality could demand so little. Would I really be fine with toasting champagne as children drown? Would I not insist on merely pushing a button to save all of Texas from imminent, consuming hellfire? How did I get to such an absurd view?

Here is the big picture. Morality either requires that we come to the aid of strangers in distress, or it doesn't. If morality doesn't require that we come to the aid of strangers, then morality is—I admit—perhaps astonishingly un-demanding. But if it does, then morality is astonishingly demanding. If we take for granted our moral reasons to save strangers are just as strong as our moral reasons to save a drowning child, then we should all be making deeply significant revisions to our lives.

You might wonder: Isn't there some middle ground? Here I acknowledge some need for caution; logical space is vast, and I can't possibly go through every way of finding a path between extremes. With that as caveat, I think the middle ground is a theoretical no man's land. If morality can demand that you sacrifice to save one child, then what about the second child, and the third? For each iteration, the reasons that saving the child was more important than your personal projects or resources will still be true. To think you're obligated to save others only up to some reasonable threshold makes sense if you can give a theoretical rationale supporting the cutoff. I doubt any theoretical rationale

DOI: 10.4324/9781003508069-25

is forthcoming. If the truth is that you can save a life by giving up your tour of baseball stadiums and you can save a life by giving up your summer activities with your children, then I don't see how there could be any prerogative for one discretionary set of values but not the other. Either impartial morality carries the day, or it doesn't.

The *very demanding view* of morality—according to which we are morally required to do what is impartially best—has excellent theoretical credentials. So too, I believe, does the *minimalist view*, according to which others have no rights whatsoever to our aid. Both of those views say our moral intuitions about aid are wildly inaccurate. And I think we should be upfront about this. Our moral intuitions, and especially our politically charged moral intuitions, are basically garbage. Intuitions are (mostly) lies we tell to get along in the world—including with ourselves. Most everything people say about morality—including cognitively sophisticated people—is based on what's socially desirable.[1] And it's very socially desirable to say that we ought to help other people. Socially desirable reported attitudes, however, are no guide at all to what's true about morality. They aren't even any guide at all to what people actually think. And almost everyone—including (not all but certainly most) philosophers—lives their lives as if they don't have obligations to strangers. So in that sense, my view is just taking seriously the way people really live as a hypothesis for what morality is really like.

Does the resulting view sound crazy? Yes. But the thing is this: Any philosophical view in the neighborhood, when applied consistently, will sound a little crazy. Having a nice fit with our ordinary moral intuitions seems like a virtue, but that virtue is an illusion.

Question 10: Is There a Moral Obligation to Beneficence?

Would I push the button to save Texas? Absolutely. It would be incomprehensible not to. But to say that something is absolutely what you should do is different from saying it's morally required. Compare: My sister is a dedicated Swiftie. Imagine she had the chance to go to a football game with Taylor. I can't conceive of her turning it down. Still, she wouldn't be morally required to go. Morality is just a part of practical normativity. On my view, it's a relatively small part, and it's good that it's a relatively small part. That doesn't prevent me from accounting for such reasons in other ways.

How should we adjudicate disagreement about what's in morality, and what's out? It won't do much good to simply issue our competing judgments—after all, we already knew that we disagreed! Instead, we need to appeal to some other aspects of our moral conceptual map and see if we get evidence one way or the other. In my initial contribution, I proposed two ways of working from premises that might be shared. No one was persuaded, and that's not surprising. If we have strong commitments about one aspect of morality, we can always revise other features of our moral concepts so that they fit together. There's nothing wrong with that.

Nevertheless, I think that our concepts of gratitude and liability to interference tell in favor of a minimal view of moral requirement. First, gratitude. My claim is that gratitude is conceptually tied to supererogation—to doing more than you have to do—and that we feel grateful to those who give aid to strangers. So, aid is more than what's morally required. Chappell denies both premises. On his view, we can be appropriately grateful to people who perform positive, required acts—like dutiful parents. On my view, the appropriateness of our gratitude should alert us to the fact that our parents were doing more than they were required to. Again, we disagree, so how can we make progress? What we need is a concept of gratitude that is (1) theoretically motivated independent of any claims about the scope of obligation and (2) empirically well supported. And my substantive proposal is that suitably independent philosophical accounts make sense of gratitude in terms of going above what morality demands.

Is my favored concept of gratitude empirically well supported? Brennan makes the point that we sometimes thank people who are only doing something they're obligated to do. He sketches an alternative theory: Maybe our gratitude isn't backward-looking (at actions surpassing obligation) but forward-looking (at trying to get people to act better). So now we have two accounts of the concept. If the attitude we're expressing is gratitude when we're trying to get people to act better, then he's got a point against me. The way to proceed is to figure out what's going on. And psychologists have some leads. For example, they distinguish gratitude (when someone does something good for you) from the related feeling of elevation (when you see someone do something good for a third party).[2] There is also evidence that gratitude and elevation are distinct emotions.[3] If it turns out that we use our concept of gratitude in Brennan's prospective way, then that's bad news for my argument. If we use praising or elevating concepts prospectively but only use gratitude retrospectively, then that's good for my side. It's an empirical question. Perhaps people do direct gratitude and not just elevation toward morally impressive but also required actions. I doubt it, because my guess is that our moral concepts are sufficiently finely grained that not all appreciative emotions will collapse into a single functional praising attitude. But I admit that it could go the other way. I'm more invested in a method of proceeding in these debates than I am in getting one or another outcome.

Second, I'll briefly consider the *argument from noninterference*. I argue that we cannot interfere with people to force them to give aid, and that if we cannot interfere to secure compliance, then there are no rights to aid and so aid is not morally required. Chappell objects that we sometimes can interfere with others to bring about that they save a life. Recall the case of the capable swimmer who could save the child, but is refusing. Chappell writes, "We may not pull a gun on them. But if they're standing on a trap door on the dock, I think you may (and should!) pull the lever to dump them into the water." My tentative diagnosis about this revised case is that people will find it acceptable to dump someone in the water because they don't really believe we have a right against being dunked in water. It's true that I *think* we have a right against the trap door being pulled. However, many people are happy to laugh along at videos of people being surreptitiously dunked in water. So I don't think everyone agrees that it's wrong. And my claim requires that the interference be morally prohibited under normal conditions. All parties to the dispute will agree that if you can get someone to do something good by performing an action that is permissible anyway, then there's no problem with that. To my mind, it is important to consider the case of waving the gun at the person, rather than pulling the trap door. If we can't threaten them with a gun, then I think that is evidence that we are not entitled to interfere with them, from which I infer (controversially, I admit) that they are not doing anything wrong.

Brennan and Chappell also pose cases in which an action is wrong, and yet third parties are not entitled to interfere.[4] It's interesting to me that these focus primarily on cases involving promises. I do find promises puzzling. I could reply in a few ways. I think sometimes promises may fail to be binding, and other times I think they may admit of enforcement. But perhaps is true that there are some genuinely required, non-binding promises. This would mean that there are two classes of moral requirement, those in which anyone can enforce the moral requirement, and the particular subclass in which these rights are limited to the recipient of a promise. I don't love this answer, but I think it could be right.[5]

Question 11: Is Effective Altruism Morally Risky?

I argued that Effective Altruism is not morally risky, except when it extends to blaming others for not being Effective Altruists. Arnold disputes both of my claims. He gives the case of the cake baker who fills their cake with razor blades. If the baker makes a cake for me, they're doing more than they have to do. But if it's filled with razor blades, then they're still doing something wrong. I accept all this, but I have a hard time seeing the analogy to the razor blades in the Effective Altruist's donation. If someone gives an organization money, then it's all cake and no razors. Arnold's example of Mark Zuckerberg is an example of someone who gives a conditional gift—offering a package of the donation plus insistence on 'market friendly' policies. The problem with the cake was that it had razor blades, and no one told me about them. To my mind, the insistence on market friendly policies doesn't sound like a razor blade, and in any case it was all part of the package upfront. If Newark didn't want a present of money +

market friendly policies, they were free to say no. And the officials who could say yes or no were democratically elected or accountable to democratically elected representatives. I still don't detect any moral risks.

On the other hand, Arnold—as well as Chappell—dispute my view that blaming people for not giving aid is risky. They both offer excellent cases, but I will focus on Arnold's. He imagines being on the phone with someone who could save his son from drowning:

> "He's right there!" I plead. "Go get him!" In response, silence. So I up the pressure: "Save him! It's a moral requirement!" Imagine you decline, politely citing your moral liberty, and then hang up. My son drowns. On Davis's analysis, *I'm* the villain in this story.

I will concede right away that I understand the villain in the story is me. I hear what Arnold is saying. But think about another case. You work your whole life so that in middle age, you'll have saved a considerable amount of money. You plan to take each of your children on a grand tour of Europe, showing them the world, and then to pay for their college education. When your oldest is graduating high school, you go to collect the money and discover—to your horror—it's gone. The investigation reveals your trusted financial advisor has taken it. He protests that he had started off pilfering some money for his son's life-saving surgery. Then, once he discovered he could skim money off your account, and realizing that other people's children also could be saved, he found he just couldn't live with himself if he stopped with his own son. He kept taking your money and kept giving it away. Always cash-strapped, he had nowhere else to turn to save each child. He continued debiting your account.

This case differs from Arnold's, in which the aggrieved parent only makes moral demands. But it won't take many steps to tie them together. If you think you'd be angry at the altruistic financial advisor, then that's evidence you think what he did is wrong. If he was wrong to take your money, that's evidence he was not entitled to interfere with you to bring about that you used your money for Effective Altruism, rather than for your own interests. If he wasn't entitled to interfere with you, then you had a right against interference. And if you had a right against being interfered with to give your money away, then (on my account of moral requirement) you were not morally required to do it.

Think about it this way. Imagine a movie told from the financial planner's point of view. He'll do anything to save his kid. My guess is that he is the hero, and that makes you the villain! Maybe this makes you rethink your moral commitments. For me, it makes me rethink my confidence in heroes. Maybe the financial planner is played by Liam Neeson. He'd not only blame you for failing to help, he'd shoot your spouse in the shoulder to make you help. For that matter, he'd burn down all of Paris to save his kid. Now, I know Arnold isn't defending this wild spin on his case. My point is that if the financial planner does all this stuff, we'll still see him as the hero. That's because our moral intuitions are way off. We've evolved to massively over-moralize other people's actions, and it's extraordinarily important that we rein in these impulses.[6] That's why I think blaming people is a morally risky business.

Question 12: Can Effective Altruism Be Part of a Meaningful Moral Life?

My answer to this question is yes, but it's up to you. You are morally free to make Effective Altruism a part of your life, or not to. Brennan reformulates the question as: "Can EA be part of a life worth living, that is, a life we would want to experience or which we would wish upon our children?"

I like this variant. I don't have children, but when I think about some young people whom I also care about, the thing I most want for them is that they live their lives by their own values, rather than someone else's values—including my own. I want for them to find projects that are fulfilling and values that repay their dedication and people they love.

Now, I realize this all sounds a bit Hallmark-esque. My point is that if I'm telling the truth, what I want for them has almost nothing to do with morality. Of course I am glad that the people

I care about are also good people, but I'm not overly exercised about it. If they happen to take a vow of poverty and live in a religious order, or if they happen to make millions and give it all away, I would be happy either way—especially if whatever they did was rewarding *to them, by their own lights*. But if, instead, they became artists or engineers or fly-fishing guides, I'd feel just as good about it. (I guess I'd be especially happy about the fly-fishing, if I'm being honest.)

Back to philosophy. I think I differ with Chappell with respect to the tie between desires and reasons. I'm on the side of thinking there must be a fairly intimate connection between a person's own mental states and what they have reason to do. In that way, I suppose I'm with Hume and against Parfit.[7] I'm sympathetic to Brennan's invocation of Jean Hampton, although there too I think Hampton puts too many moral demands on other people (though from a very different direction).

Finally, Chappell and Arnold bring up the possibility that you could try to shape your own attitudes to include a strong final desire to do what's morally most important. Arnold suggests a path of "empathetic broadening," whereby one cultivates a concern for others, including strangers. Imagine the baseball fan could take a pill to shift his preferences to be more beneficent. "Should he take the pill?" Arnold asks.

My answer will be sincere but unsatisfying: It depends. I would want to ask a few questions of our imagined baseball fan. Is taking the pill part of a suite of activities to realize his own values? Would parts of his identity fit together better—would he be a more coherent person to himself and others—if he made the change? How would he look back on this turning point in the story of his life? Does he want to be the more beneficent version of himself? Many of us have a 'two wolves' problem in us—with different parts of us wanting conflicting things. Is it better to spend more time with family, or throw oneself into making the world better? Should I be trying to make it in philosophy, or should I be out wandering the prairie? I can't settle questions like this for you, and you can't answer them for me. If we imagine our hypothetical baseball fan as not just a baseball fan but also a person, then what I should do is respect him.

Notes

1 On the effects of philosophical training on intuitions, see Maćkiewicz, Kuś, and Hensel 2023.
2 Algoe and Haidt 2009; McCullough, Kimeldorf, and Cohen 2008.
3 Siegel, Thomson, and Navarro 2014.
4 Arnold also makes some nice conceptual objections to this argument, and I would like to think more about them, but for space I'm omitting them here and focusing on a version of his complaint in the next section.
5 Such a disjunctive view might sound *ad hoc*, but Setiya (2022) says some things to allay this worry.
6 Fiske 2014.
7 Street 2009.

BIBLIOGRAPHY

Achen, Christopher and Larry Bartels. 2016. *Democracy for Realists*. Princeton University Press.
Alcott, Hunt, Giovanni Montanari, Bora Ozaltun, and Brandan Tan. 2023. "An Economic View of Corporate Social Impact," National Bureau of Economic Research, Working Paper No. 31803. Available at : https://www.nber.org/papers/w31803?fbclid=IwAR1FcWAx_0Oai kTX19FKhcphecP_UdJwRCFoyEe0JnsTvAbzAhwtZd9TI_E#fromrss
Algoe, Sara B. and Jonathan Haidt. 2009. "Witnessing Excellence in Action: The 'Other-Praising' Emotions of Elevation, Gratitude, and Admiration," *The Journal of Positive Psychology* 4: 105–127.
Aos, Steve, Roxanne Lieb, Jim Mayfield, et al. 2004. "Benefits and Costs of Prevention and Early Intervention Programs for Youth," Washington State Institute for Public Policy. Available at: www.wsipp.wa.gov/rptfiles/04-07-3901b.pdf
Arnold, Denis G. and Norman E. Bowie. 2003. "Sweatshops and Respect for Persons," *Business Ethics Quarterly* 13: 221–242.
Arnold, Samuel. 2016. "Socialism," *Internet Encyclopedia of Philosophy*. Available at: iep.utm.edu
Arnold, Samuel, 2017. "Capitalism, Class Conflict, and Domination," *Socialism and Democracy* 31: 106–117.
Arnold, Samuel. 2021. "Contesting the Work-Spend Cycle: The Liberal Egalitarian Case Against Consumerism," in *The Politics and Ethics of Contemporary Work*, ed. Keith Breen and Jean-Philippe Deranty. Routledge.
Arnold, Samuel. 2022a. "Socialisms," in *The Routledge Handbook of Philosophy, Politics, and Economics*, ed. C. M. Melenovsky, 276–288. Routledge.
Arnold, Samuel. 2022b. "No Community Without Socialism: Why Liberal Egalitarianism Is Not Enough," *Philosophical Topics* 48: 1–22.

Arntzenius, Frank, Adam Elga, and John Hawthorne. 2004. "Bayesianism, Infinite Decisions, and Binding," *Mind* 113: 251–283.

Barrett, J. 2022. "Social Beneficence," GPI Working Paper No. 11–2022. Available at: https://philpapers.org/rec/BARSBO-2

Bazerman, Max H. and Ann E. Tenbrunsel. 2011. *Blind Spots: Why We Fail to Do What's Right and What to Do About It*. Princeton University Press.

Beerbohm, Eric. 2016. "The Free-Provider Problem: Private Provision of Public Responsibilities," in *Philanthropy in Democratic Societies*, eds. Rob Reich, Lucy Bernholz, and Chiara Cordelli, 207–225. University of Chicago Press.

Bhagwat, Y., N. L. Warren, J. T. Beck, and G. F. Watson, 2020. "Corporate Sociopolitical Activism and Firm Value," *Journal of Marketing* 84: 1–21.

Bisgaard. Martin and Rune Slothuus. 2018. "Partisan Elites as Culprits? How Party Cues Shape Partisan Perceptual Gaps," *American Journal of Political Science* 62: 456–469.

Bostrom, Nick. 2003. "Transhumanist Values," in *Ethical Issues for the 21st Century*, ed. Frederick Adams, Philosophical Documentation Center. Available at https://nickbostrom.com/ethics/values

Bostrom, Nick. 2008. "Letter from Utopia," *Studies in Ethics, Law, and Technology* 2: 1–7.

Bostrom, Nick. 2013. "Existential Risk Prevention as a Global Priority," *Global Policy* 4: 15–31.

Brennan, Jason. 2009. "Polluting the Polls: When Citizens Should Not Vote," *Australasian Journal of Philosophy* 87: 535–549.

Brennan, Jason. 2011. *The Ethics of Voting*. Princeton University Press.

Brennan, Jason. 2014. *Why Not Capitalism?* Routledge.

Brennan, Jason. 2016. *Against Democracy*. Princeton University Press.

Brennan, Jason. 2020. *Why It's OK to Want to Be Rich*. Routledge.

Brennan, Jason and Philip Magness. 2019. *Cracks in the Ivory Tower*. Oxford University Press.

Buss, Sarah. 2006. "Needs (Someone Else's), Projects (My Own), and Reasons," *The Journal of Philosophy* 103: 373–402.

Callahan, David. 2017. *The Givers*. Vintage Books.

Caplan, Bryan. 2007. *The Myth of the Rational Voter*. Princeton University Press.

Carens, Joseph. 2015. *The Ethics of Immigration*. Oxford University Press.

Caviola, L., S. Schubert, and J. D. Greene. 2021. "The Psychology of (In) Effective Altruism," *Trends in Cognitive Sciences* 25: 596–607.

Chappell, Richard Yetter. 2017. "Rethinking the Asymmetry," *Canadian Journal of Philosophy* 47: 167–177.

Chappell, Richard Yetter. 2019a. "Willpower Satisficing," *Noûs* 53: 251–265.

Chappell, Richard Yetter. 2019b. "Overriding Virtue," in *Effective Altruism: Philosophical Issues*, eds. Hilary Greaves and Theron Pummer, 218–226. Oxford University Press.

Chappell, Richard Yetter. Forthcoming. "Consequentialism: Core and Expansion," in *The Oxford Handbook of Normative Ethics*, eds. D. Copp, C. Rosati, and T. Rulli. Oxford University Press.

Chappell, Richard and Helen Yetter-Chappell. 2016. "Virtue and Salience," *Australasian Journal of Philosophy* 94: 449–463.

Chappell, R. Y., D. Meissner, and W. MacAskill. 2023. *An Introduction to Utilitarianism*. Available at: www.utilitarianism.net.

Claassen, Ryan L. et al. 2021. "Which Party Represents My Group? The Group Foundations of Partisan Choice and Polarization," *Political Behavior* 43: 615–636.

Cohen, G. A. 1995. *Self-Ownership, Freedom, and Equality*. Cambridge University Press.

Cordelli, Chiara. 2020. *The Privatized State*. Princeton University Press.

Coyne, Christopher. 2013. *Doing Bad by Doing Good*. Stanford University Press.

Darwall, Stephen. 2006. *The Second-Person Standpoint: Morality, Respect, and Accountability*. Harvard University Press.

Davis, Ryan W. 2019. "What Must Good Samaritans Do? Skepticism About the Political Enforceability of Duties to Aid," *Public Affairs Quarterly* 33: 41–64.

Davis, Ryan W. and Jessica Preece. 2022. "Individual Valuing of Social Equality in Political and Personal Relationships," *Review of Philosophy and Psychology* 13: 177–196.

Deaton, Angus. 2013. *The Great Escape*. Princeton University Press.

Deci, Edward L. and Richard M. Ryan. 2014. "Autonomy and Need Satisfaction in Close Relationships: Relationships Motivation Theory," in *Human Motivation and Interpersonal Relationships: Theory, Research, and Applications*, ed. Netta Weinstein, 53–73. Springer.

Deneen, Patrick. 2023. *Regime Change: Towards a Postliberal Future*. Sentinel.

Deri, Sebastian, Daniel H. Stein, and Vanessa K. Bohns. 2019. "With a Little Help from My Friends (and Strangers): Closeness as a Moderator of the Underestimation-of-Compliance Effect," *Journal of Experimental Social Psychology* 82: 6–15.

Desvousges, William H., F. Reed Johnson, Richard W. Dunford, Kevin J. Boyle, Sara P. Hudson, and K. Nicole Wilson. 2010. *Measuring Nonuse Damages Using Contingent Valuation: An Experimental Evaluation of Accuracy*. RTI Press.

Dias, Nicholas and Yphtach Lelkes. 2022. "The Nature of Affective Polarization: Disentangling Policy Disagreement from Partisan Identity," *American Journal of Political Science* 66: 775–790.

Dorsey, Dale. 2009. "Aggregation, Partiality, and the Strong Beneficence Principle," *Philosophical Studies* 146: 139–157.

Douthat, Ross. 2018. "The Rise of Woke Capital," *New York Times*, February 28, 2018. https://www.nytimes.com/2018/02/28/opinion/corporate-america-activism.html (accessed August 27, 2023).

Duda, Roman. 2017. "Building Effective Altruism." Available at: https://80000hours.org/problem-profiles/promoting-effective-altruism/#:~:text=Effective%20altruism%20is%20about%20using,best%20opportunities%20for%20doing%20good

Eikenberry, Angela M. and Roseanne Marie Mirabella. 2018. "Extreme Philanthropy: Philanthrocapitalism, Effective Altruism, and the Discourse of Neoliberalism," *PS: Political Science & Politics* 51: 43–47.

Feinberg, Joel. 1974. "On the Rights of Animals and Future Generations," in *Philosophy and Environmental Crisis*, ed. William Blackstone, 43–68. University of Georgia Press.

Fink, Larry. 2018. "Sense of Purpose," Available at: https://www.blackrock.com/corporate/investor-relations/2018-larry-fink-ceo-letter (accessed August 27, 2023).

Fiske, Alan Page. 2014. *Virtuous Violence*. Illustrated edition. Cambridge University Press.

Flanigan, Jessica. 2019. "Duty and Enforcement," *Journal of Political Philosophy* 27: 341–362.

Flanigan, Jessica and Christopher Freiman. 2022. "Wealth Without Limits: In Defense of Billionaires," *Ethical Theory and Moral Practice* 25: 755–775.

Flynn, D. J., Brendan Nyhan, and Jason Reifler. 2017. "The Nature and Origins of Misperceptions: Understanding False and Unsupported Beliefs About Politics," *Political Psychology* 38: 127–150.

Frankfurt, Harry G. 1988. *The Importance of What We Care About: Philosophical Essays*. Cambridge University Press.

Frederiksen, Kristian Vrede Skaaning. 2022. "Does Competence Make Citizens Tolerate Undemocratic Behavior?," *American Political Science Review* 116: 1147–1153.

Frick, J. 2020. "Conditional Reasons and the Procreation Asymmetry," *Philosophical Perspectives* 34 (1): 53–87.

Gabriel, Iason. 2015. "Effective Altruism and Its Critics," *Journal of Applied Philosophy* 34: 457–473.

Gauthier, David. 1987. *Morals by Agreement*. Oxford University Press.

Gert, Joshua. 2007. "Normative Strength and the Balance of Reasons," *The Philosophical Review* 116: 533–562.

Gibbard, Allan. 1984. "Utilitarianism and Human Rights," *Social Philosophy and Policy* 1 (2): 92–102.
Grodeck, Ben and Philipp Schoenegger. 2023. "Demanding the Morally Demanding: Experimental Evidence on the Effects of Moral Arguments and Moral Demandingness on Charitable Giving," *Journal of Behavioral and Experimental Economics* 103: 101988. https://doi.org/10.1016/j.socec.2023.101988.
Hampton, Jean. 1987. *Hobbes and the Social Contract Tradition.* Cambridge University Press.
Hampton, Jean. 1993. "Selflessness and the Loss of Self," *Social Philosophy and Policy* 10: 135–165.
Hare, Caspar. 2016. "Should We Wish Well to All?," *Philosophical Review* 125: 451–472.
Harman, Elizabeth. 2016. "Morally Permissible Moral Mistakes," *Ethics* 126 (2): 366–393.
Hasnas, John. 2013. "Whither Stakeholder Theory? A Guide for the Perplexed Revisited," *Journal of Business Ethics* 112: 47–57.
Hasner, Matthew. 2014. "Imperfect Aiding," In *The Cambridge Companion to Life and Death*, ed. Steven Luper, 300–315. Cambridge University Press.
Hayek, Friedrich A. von. 1945. "The Use of Knowledge in Society," *American Economic Review* 35: 519–530.
Henderson, Rebecca. 2020. *Reimagining Capitalism in a World on Fire.* PublicAffairs.
Herman, Barbara. 2001. "The Scope of Moral Requirement," *Philosophy & Public Affairs* 30: 227–256.
Herman, Barbara. 2012. "Being Helped and Being Grateful: Imperfect Duties, the Ethics of Possession, and the Unity of Morality," *The Journal of Philosophy* 109: 391–411.
Hidalgo, Javier. 2018. *Unjust Borders: Individuals and the Ethics of Immigration.* Routledge.
Hiller, Avram and Ali Hasan. Forthcoming. "How to Save Pascal (and Ourselves) from the Mugger," *Dialogue*.
Hobbes, Thomas. 1651. *Leviathan.* Available at: www.gutenberg.org/files/3207/3207-h/3207-h.htm
Horton, Joe. 2017. "The All or Nothing Problem," *The Journal of Philosophy* 114: 94–104.
Huemer, Michael. 2004. "America's Unjust Drug War," in *The New Prohibition*, ed. Bill Masters, 133–144. Accurate Press.
Huemer, Michael. 2010. "Is There a Right to Immigrate?," *Social Theory and Practice* 36: 429–461.
Huemer, Michael. 2012. *The Problem of Political Authority: An Examination of the Right to Coerce and the Duty to Obey.* Palgrave Macmillan.

Hume, David. 1739. *A Treatise of Human Nature*. Available at: www.gutenberg.org/files/4705/4705-h/4705-h.htm

Hursthouse, Rosalind. 1995. *On Virtue Ethics*. Oxford University Press.

Hussain, Waheed. 2012. "Is Ethical Consumerism an Impermissible Form of Vigilantism?," *Philosophy & Public Affairs* 40: 111–142.

Iyengar, Shanto and Sean J. Westwood. 2015. "Fear and Loathing Across Party Lines: New Evidence on Group Polarization," *American Journal of Political Science* 59: 690–707.

Jackson, Frank. 1991. "Decision-Theoretic Consequentialism and the Nearest and Dearest Objection," *Ethics* 101: 461–482.

Jones, Charles. 2013. "The Human Right to Subsistence," *Journal of Applied Philosophy* 30 (1): 57–72.

Kalmoe, Nathan and Lilliana Mason. 2022. *Radical American Partisanship: Mapping Violent Hostility, Its Causes, and the Consequences for Democracy*. University of Chicago Press.

Karpowitz, Christopher F., Lisa Argyle, and Ryan W. Davis. n.d. "The Good Deliberative Citizen: How Partisanship Shapes Citizenship Norms About Deliberative Behaviors," unpublished manuscript.

Kavka, Gregory. 2021. *Hobbesian Moral and Political Theory*. Princeton University Press.

Kinder, Donald and Nathan Kalmoe. 2017. *Neither Liberal Nor Conservative: Ideological Innocence in the American Public*. University of Chicago Press.

Korsgaard, Christine M. 2009. *Self-Constitution: Agency, Identity, and Integrity*. Oxford University Press.

Kosonen, P. 2023. "Tiny Probabilities and the Value of the Far Future," GPI Working Paper No. 1–2023.

Krishnarajan, Suthan. 2022 "Rationalizing Democracy: The Perceptual Bias and (Un)Democratic Behavior," *American Political Science Review* 177: 474–496.

Landwehr, Claudia and Philipp Harms. 2020. "Preferences for Referenda: Intrinsic or Instrumental? Evidence from a Survey Experiment," *Political Studies* 68: 875–894.

Law, Kyle Fiore, Dylan Campbell, and Brendan Gaesser. 2022. "Biased Benevolence: The Perceived Morality of Effective Altruism Across Social Distance," *Personality and Social Psychology Bulletin* 48: 426–444.

Lechterman, Theodore. 2022. *The Tyranny of Generosity*. Oxford University Press.

Lechterman, Theodore, Ryan Jenkins, and Bradley Strawser. 2024. "#StopHateForProfit and the Ethics of Boycotting by Corporations," *Journal of Business Ethics* 191 (1): 77–91. DOI: 10.1007/s10551-023-05415-y.

Lewis, Michael. 2023. *Going Infinite: The Rise and Fall of a New Tycoon*. W.W. Norton & Co.

Liberto, Hallie. 2017. "The Problem with Sexual Promises," *Ethics* 127: 383–414.
Logins, Artūrs. 2016. "Save the Children!" *Analysis* 76: 418–422.
MacAskill, William. 2016. *Doing Good Better*. Avery Press.
MacAskill, William. 2022. *What We Owe the Future*. Basic Books.
MacAskill, W., K. Bykvist, and T. Ord. 2020. *Moral Uncertainty*. Oxford University Press.
MacAskill, William, Benjamin Todd, and Robert Wiblin. 2015. "Can You Guess Which Government Programs Work? Most People Can't," *Vox*, August 17, 2015. https://www.vox.com/2015/8/13/9148123/quiz-which-programs-work.
MacFarquhar, Larissa. 2016. *Strangers Drowning: Impossible Idealism, Drastic Choices, and the Urge to Help*, Reprint edition. Penguin Books.
Machery, Edward. 2019. *Philosophy within Its Proper Bounds*. Oxford University Press.
Maćkiewicz, Bartosz, Katarzyna Kuś, and Witold M. Hensel. 2023. "The Influence of Philosophical Training on the Evaluation of Philosophical Cases: A Controlled Longitudinal Study," *Synthese* 202: https://doi.org/10.1007/s11229-023-04316-x.
MacNamara, Coleen. 2013. "'Screw You!' & 'Thank You.'" *Philosophical Studies* 165: 893–914.
Maier, Maximilia, Lucius Caviola, Stefan Schubert, and Adam Harris. 2023. "Investigating (Sequential) Unit Asking: An Unsuccessful Quest for Scope Sensitivity in Willingness to Donate Judgments," *Behavioral Decision Making* 36. https://doi.org/10.1002/bdm.2335.
Martela, Frank and Richard M. Ryan. 2016. "The Benefits of Benevolence: Basic Psychological Needs, Beneficence, and the Enhancement of Well-Being," *Journal of Personality* 84: 750–764.
Masconale, Saura and Simone Sepe. 2022. "Citizen Corp.: Corporate Activism and Democracy," *Washington University Law Review* 100: 257–325.
Mason, Lilliana. 2017. *Uncivil Agreement: How Politics Became Our Identity*. University of Chicago Press.
Mason, Lilliana. 2018. "Ideologues Without Issues: The Polarizing Consequences of Ideological Identities," *Public Opinion Quarterly* 82: 280–301.
McCullough, Michael E., Shelley D. Kilpatrick, Robert A. Emmons, and David B. Larson. 2001. "Is Gratitude a Moral Affect?," *Psychological Bulletin* 127: 249–266.
McCullough, Michael E., Marcia B. Kimeldorf, and Adam D. Cohen. 2008. "An Adaptation for Altruism: The Social Causes, Social Effects, and Social Evolution of Gratitude," *Current Directions in Psychological Science* 17: 281–285.

McMahan, J. 2013. "Causing People to Exist and Saving People's Lives," *The Journal of Ethics* 17: 5–35.

McPherson, Tristram. 2018. "Authoritatively Normative Concepts," in *Oxford Studies in Metaethics 13*, ed. Russ Shafer-Landau. Oxford University Press. https://doi.org/10.1093/oso/9780198823841.003.0012.

Mill, John Stuart. 1863. *Utilitarianism*. Available at: https://www.utilitarianism.net/books/utilitarianism-john-stuart-mill/2/

Minson, J. A. and B. Monin. 2012. "Do-Gooder Derogation: Disparaging Morally Motivated Minorities to Defuse Anticipated Reproach," *Social Psychological and Personality Science* 3: 200–207.

Moller, Dan. 2019. *Governing Least: A New England Libertarianism*. Oxford University Press.

Muñoz, Daniel. 2021. "Exploitation and Effective Altruism," *Politics, Philosophy & Economics* 20: 409–423.

Nagel, Thomas. 1971. "The Absurd," *The Journal of Philosophy* 68: 716–727.

Nagel, Thomas. 1978. *The Possibility of Altruism*. Princeton University Press.

Narveson, Jan. 1973. "Moral Problems of Population," *The Monist* 57: 62–86.

National Philanthropic Trust. 2023. "Charitable Giving Statistics." Available at: https://www.nptrust.org/philanthropic-resources/charitable-giving-statistics/

Nickel, Patricia Mooney and Angela M. Eikenberry. 2009. "A Critique of the Discourse of Marketized Philanthropy," *American Behavioral Scientist* 52: 974–989.

Noggle, Robert. 2009. "Give Till It Hurts? Beneficence, Imperfect Duties, and a Moderate Response to the Aid Question," *Journal of Social Philosophy* 40: 1–16.

Nozick, Robert. 1974. *Anarchy, State, and Utopia*. Basic Books.

Ord, Toby. 2021. *The Precipice*. Hachette.

Parfit, Derek. 1978. "Innumerate Ethics," *Philosophy and Public Affairs* 7: 285–301.

Parfit, Derek. 1984. *Reasons and Persons*. Oxford University Press.

Parfit, Derek. 1987. *Reasons and Persons*. Revised edition. Clarendon Press.

Paul, Darel. 2022. "The Puzzle of Woke Capital," *American Affairs*. Available at: https://americanaffairsjournal.org/2022/08/the-puzzle-of-woke-capital/ (accessed August 27, 2023).

Petrosino, Anthony, Carolyn Turpin-Petrosino, Meghan E. Hollis-Peel, and Julia G. Lavenberg. 2013. "'Scared Straight' and Other Juvenile Awareness Programs for Preventing Juvenile Delinquency," *The Cochrane Database of Systematic Reviews* 4: CD002796. doi:10.1002/14651858.CD002796.pub2

Pettit, Philip. 1991. "Consequentialism," in *A Companion to Ethics*, ed. Peter Singer, 230–240. Blackwell.
Pew Research. 2020. "Most Democrats Would Not Consider Dating a Trump Voter." Available at: https://www.pewresearch.org/short-reads/2020/04/24/most-democrats-who-are-looking-for-a-relationship-would-not-consider-dating-a-trump-voter/ (accessed August 27, 2023).
Pogge, Thomas. 2007. *World Poverty and Human Rights*, 2nd edition. Polity.
Politifact. 2021. Fact checks. Available at: https://www.politifact.com/factchecks/2021/nov/02/viral-image/confiscating-us-billionaires-wealth-would-run-us-g/
Pollock, John. 1983. "How Do You Maximize Expectation Value?" *Noûs* 17: 409–421.
Pummer, Theron. 2016. "Whether and Where to Give," *Philosophy & Public Affairs* 44: 77–95.
Pummer, Theron. 2023. *The Rules of Rescue*. Oxford University Press.
Railton, Peter. 1984. "Alienation, Consequentialism, and the Demands of Morality," *Philosophy and Public Affairs* 13 (2): 134–171.
Rawls, John. 1999. *A Theory of Justice*, revised edition. Harvard University Press.
Reich, Rob. 2013. "What Are Foundations For?," *Boston Review*. Available at: https://www.bostonreview.net/forum/foundations-philanthropy-democracy/
Reich, Rob. 2018. *Just Giving: Why Philanthropy Is Failing Democracy and How It Can Do Better*. Princeton University Press.
Rekker, Roderik and Eelco Harteveld. 2022. "Understanding Factual Belief Polarization: The Role of Trust, Political Sophistication, and Affective Polarization," *Acta Politica*. Available at: https://doi.org/10.1057/s41269-022-00265-4
Ross, W. D. 1930. *The Right and the Good*. Oxford University Press.
Ruckelshaus, Jay. 2022. "What Kind of Identity Is Partisan Identity? 'Social' versus 'Political' Partisanship in Divided Democracies," *American Political Science Review* 116: 1477–1489.
Ryan, Richard M. et al. 2021. "Building a Science of Motivated Persons: Self-Determination Theory's Empirical Approach to Human Experience and the Regulation of Behavior," *Motivation Science* 7: 97–110.
Sabato, Hagit and Sapir Bar-Ilan. 2023. "Pleasure or Meaning: Subjective Well-Being Orientations and the Willingness to Help Close Versus Distant Others," *Journal of Happiness Studies* 24: 2013–2037.
Sachs, Ben. 2019. "Demanding the Demanding," in *Effective Altruism: Philosophical Issues*, eds. Hilary Greaves and Theron Pummer. Oxford University Press.
Saunders-Hastings, Emma. 2022a. "Economic Power and Democratic Forbearance," in *Wealth and Power: Philosophical Perspectives*,

eds. Michael Bennett, Huub Brouwer, and Rutger Classen, 186–205. Routledge.

Saunders-Hastings, Emma. 2022b. *Private Virtues, Public Vices: Philanthropy and Democratic Equality*. University of Chicago Press.

Scheffler, Samuel. 2012. *Equality and Tradition: Questions of Value in Moral and Political Theory*. Oxford University Press.

Schmidtz, David. 1995. *Rational Choice and Moral Agency*. Princeton University Press.

Schmidtz, David. 2006. *The Elements of Justice*. Cambridge University Press.

Schwitzgebel, E. 2019. "Aiming for Moral Mediocrity," *Res Philosophica* 96: 347–368.

Setiya, Kieren. 2014. "Love and the Value of a Life," *The Philosophical Review* 123: 251–280.

Setiya, Kieren. 2022. "What Is Morality?," *Philosophical Studies* 179: 1113–1133. https://doi.org/10.1007/s11098-021-01689-y.

Shue, Henry. 1996. *Basic Rights: Subsistence, Affluence, and U.S. Foreign Policy*. Princeton: University Press.

Siegel, Jason T., Andrew L. Thomson, and Mario A. Navarro. 2014. "Experimentally Distinguishing Elevation from Gratitude: Oh, the Morality," *The Journal of Positive Psychology* 9: 414–427.

Simler, Kevin and Robin Hanson. 2018. *The Elephant in the Brain*. Oxford University Press.

Singer, Peter. 1972. "Famine, Affluence, and Morality," *Philosophy & Public Affairs* 1: 229–243.

Singer, Peter. 2009. *The Life You Can Save: Acting Now to Stop World Poverty*. Random House.

Smart, J. J. C. and Bernard Williams. 1973. *Utilitarianism: For and Against*. Cambridge University Press.

Smith, Michael. 2011. "Deontological Moral Obligations and Non-Welfarist Agent-Relative Values," *Ratio* 24: 351–363.

Street, Sharon. 2009. "In Defense of Future Tuesday Indifference: Ideally Coherent Eccentrics and the Contingency of What Matters," *Philosophical Issues* 19: 273–298.

Syme, Timothy. 2019. "Charity vs. Revolution: Effective Altruism and the Systemic Change Objection," *Ethical Theory and Moral Practice* 22: 93–120.

Talisse, Robert. 2019. *Overdoing Democracy*. Oxford University Press.

Tan, Kok-Chor. 2004. "Justice and Personal Pursuits," *The Journal of Philosophy* 101(7): 331–362.

Taurek, John. 1977. "Should the Numbers Count?," *Philosophy and Public Affairs* 6: 293–316.

Temkin, Larry. 2004. "Thinking about the Needy, Justice, and International Organizations," *The Journal of Ethics* 8 (4): 349–395.
Temkin, Larry. 2022. *Being Good in a World of Need*. Oxford University Press.
Thomson, Judith Jarvis. 1971. "A Defense of Abortion," *Philosophy and Public Affairs* 1: 47–66.
Timmerman, Travis. 2015. "Sometimes There Is Nothing Wrong with Letting a Child Drown," *Analysis* 75: 204–212.
Torres, Émile P. 2023a. "The Ethics of Human Extinction," *Aeon Magazine*. https://aeon.co/essays/what-are-the-moral-implications-of-humanity-going-extinct (accessed 24 August 2023).
Torres, Émile P. 2023b. *Were the Great Tragedies of History 'Mere Ripples'? The Case Against Longtermism*. Available at: https://www.xriskology.com/mini-book (accessed November 25, 2023).
Tosi, Justin and Brandon Warmke. 2024. *Why It's OK to Mind Your Own Business*. Routledge.
Tucker, Chris. 2022a. "Too Far Beyond the Call of Duty: Moral Rationalism and Weighing Reasons," *Philosophical Studies* 179: 229–252.
Tucker, Chris. 2022b. "The Dual Scale Model of Weighing Reasons," *Noûs* 56: 366–392.
Tversky, Amos, and Daniel Kahneman. 1974. "Judgment under Uncertainty: Heuristics and Biases: Biases in Judgments Reveal Some Heuristics of Thinking Under Uncertainty," *Science* 185 (4157): 1124–1131.
Unger, Peter. 1996. *Living High While Letting Die*. Oxford University Press.
Vegetti, Federico and Moreno Mancosu. 2020. "The Impact of Political Sophistication and Motivated Reasoning on Misinformation," *Political Communication* 37: 678–695.
Velleman, J. David. 1999. "Love as a Moral Emotion," *Ethics* 109: 338–374.
Wallace, R. Jay. 2004. "The Rightness of Acts and the Goodness of Lives," in *Reason and Value: Themes from the Moral Philosophy of Joseph Raz*, eds. R. Jay Wallace, Philip Pettit, Samuel Scheffler, and Michael Smith, 385–411. Clarendon Press.
Weinstein, Netta and Richard M. Ryan. 2010. "When Helping Helps: Autonomous Motivation for Prosocial Behavior and Its Influence on Well-Being for the Helper and Recipient," *Journal of Personality and Social Psychology* 98: 222–244.
West, Emily and Shanto Iyengar. 2022. "Partisanship as a Social Identity: Implications for Polarization," *Political Behavior* 44: 807–838.

INDEX

Note: Page numbers followed by "n" denote endnotes.

acting ethically *see* virtue signals
actions: moral 120, 130, 139, 185; supererogatory 139, 182, 189, 209
activism (corporate) 18–20
advertisement (of beneficence) 120–1, 129, 140
agential assessment *see* moral agents
aid illusion 108
altruism: effective altruism (theory) 1, 64, 96, 156; pathological altruism 149, 156, 194; and well-being 149, 163; *see also* Effective Altruism (EA)
animal welfare 86–7
anonymous donations 120–1, 129
argument from noninterference 166–7, 198, 210–11
Arnold, Laura 15

bank robbery example (justice) 9–10, 43
baseball fan example (meaningful life) 174, 176, 184–6, 208, 214
benchmark biases 187
beneficence: assessing 97, 118–19; versus benevolence 72–3, 98; and business (corporate) 19, 83–5, 106; conflict with political values 49–51; convenient 94; as donor constraint 8–9; at expense of other values (motivations) 151–3; versus fighting injustice 116–17; importance of 116, 139, 145–6, 151–2; moral reasons of 98–9, 101, 115, 153–4; and moral theories 90–1; as moral value 116, 137–9, 151–2; obligation to 66–7, 71, 82, 100, 109–11, 163–8; in other domains 68–71, 91; philosophical theories 187–8; in professions 68–9; versus sacrifice 116, 140; unintentional 94; and utilitarianism 91, 115, 138
beneficentrism 90–1
beneficiaries (of charities) 70, 92–3
benevolence: versus beneficence 72–3, 98; impartial 131, 156; as master virtue 131–2, 155
biases: benchmark biases 187; social desirability 101, 208
Bloomberg, Mike 14
Bostrom, Nick: "Maxipok" (moral rule) 133–6
branding (consumer goods) 75

burdens 93–4, 194
burning house example (moral risks) 171–2
businesses: accountability (value assessment) 69–70, 106; and beneficence 83–5, 106, 192; effect on background conditions 73–5; investment in development 108–9; market failure 74, 106; political stance of 32–4; as producers of negative externalities 86, 106; profitability and social good 85–6; purpose of 19; social businesses 69–70, 83, 93; *see also* corporate activism; Corporate Social Responsibility (CSR)
business ethics (theories of) 103–6

cake baker (razor blades) example (moral risks) 182–3, 211
Callahan, David 17
capital flight (threat of) 22, 56
capitalism 84, 88n4, 171
Carens, Joseph 12
Catastrophically Bad Lifeguard example *see* lifeguard example
cause prioritization 201–2, 204
Caviola, Lucius: *"The Psychology of (In)Effective Altruism"* 95
character traits 131–2
charitable giving 77–8, 95, 107
charities: accountability (value assessment) 69–71, 92; central problems of 69, 70, 99; donors versus beneficiaries 70, 92–3, 99; effective 74, 77, 92, 95; Effective Altruism charities 80, 107, 191; and elimination of global poverty 109; GiveWell 92, 93; mission creep 191
charity work (motives for) 76–9
Cohen, G. A. 143
commonsense morality 66–7, 188
communication (signaling) 75–6
community (and donor discretion) 27–8, 68, 79

complacency (moral) *see* moral complacency
complicity: in democratic societies 10–12, 28–9; everyday 10–12, 28, 46; and inaction 28; in state's injustices 10–12, 28, 37–8, 45–7, 51–2; of tax-payers/voters 29
concerns priorities (objective versus subjective) 203–5
confirmation bias 77
consequentialists 36, 131, 132–5, 147
consumer goods (and global economies) 74–5, 91–2
consumption (beneficence of) 84
convenient beneficence 94
Cordelli, Chiara 10, 13, 51
corporate activism 18–20; capital flight (threat of) 22, 56; and democracy 22; influence of 56; and overpoliticization 23–4, 32, 41, 47; and political equality 21–3, 25; and political neutrality 41–2; and public opinion 22–3; social benefits of 41; veto (power of) 22, 55–7; "woke capitalism" 20; versus workers' interests 34
corporate beneficence 19
corporate political activism 24
Corporate Social Responsibility (CSR) 19, 85, 104–6
corporations *see* businesses
cultural influence 31, 39
customers (of social businesses) 70

Deaton, Angus 107–8
Default View (of philanthropy) 7–8, 13, 43
defeasible claims 51–2
democracy: and corporate activism 21, 22; and egalitarian processes 16–18, 21, 53; and (elite) philanthropy 14–18, 30–1, 39–41, 53–4, 170; and everyday complicity 10–12, 28–9, 46; failure

of governments 31–2; versus plutocracy 16, 54–5, 57; problems of 47; and public opinion 22–3; versus undemocratic philanthropy 30–1
democratic theory 62–3
democratic voting 32
deontic moral lien: and donor discretion 9–10, 13, 50–1, 53; and everyday complicity 10–11
deontology (and beneficentrism) 90–1
development (and global poverty) 108–9
Diallo, Ray 18
distributive injustice 11–12
'do-gooder derogation' 121
donor discretion: anonymous donations 120–1, 129; beneficence as constraint 8–9; choice of charities 77–8; community consideration 27–8; confirmation bias 77; considerations 28, 68, 77–8; and deontic moral lien 9–10, 13, 50–1, 53; family and fraternity 28; giving democratically 13; giving justly and well 8, 13; (in)justice as constraint 9, 36–7, 44; and moral freedom 8–9, 163; "scope insensitivity" 77; splitting donations 95; *see also* looking good versus doing good
Douthat, Ross 20
drowning child example 95, 120, 165–6, 183–4, 188–9; capable swimmer 210–11; gratitude (moral view of) 199; moral complacency 200
Duda, Roman: *80,00 Hours* 64

earning to give 69, 71
economics games 78–9
Effective Altruism (EA) 156; burning house example 171–2; charities 80, 107, 191; defining 1, 62, 98; duty to help 66–7, 71, 83; *earning to give* 69, 71; growth of 96; ideas and questions 65; intentions (judgment of) 73, 156; messaging 80; Minimally Effective Altruism 67; and moral life 184–6, 192–3, 203, 213–14; moral perspective of 89–90, 184–6; moral prioritization of 115, 148–9; moral risks of 169–73, 183–4, 190, 200, 211–13; and personal relationships 148; and philanthropy 27, 44; and religion 192; as substantive theory 106; as undemocratic 170; and utilitarianism 63–4, 66, 89, 98, 170–1; *see also* moral requirement (to help)
effective altruism (theory) 1, 64, 96, 156
effective charities *see* charities
egalitarian processes: and democracy 16–18, 21, 53; and elite philanthropy 17–18, 21, 53
elevation (moral view of) 210
elite philanthropy 14–16, 30; versus government spending 31; and influence 21–2, 38–9, 54; political equality 17–18, 53, 55; procedural unfairness of 15, 38, 54; undemocratic nature of 14–18, 30–1, 53–4
Emerson, Ralph Waldo 100
empathetic broadening 214
Environmental, Social, and Governance (ESG) 85, 104–6
equality (political) 17–18, 53, 55
existential risk *see* extinction (of humanity)
externalities, negative 86–7
extinction (of humanity) 125–6, 133–5, 157–8, 159n7

Index

family and fraternity 28; *see also* moral requirement (to help)
fanaticism (moral) *see* moral fanaticism
financial advisor example (moral risks) 212–13
Flanigan, Jessica 32
"flawed realization" 135
Freiman, Christopher 32, 67–8
Friedman, Milton 85
frivolous personal consumption 8
future generations 116; concern for 123–7, 132–5, 141–3; negative externalities on 87; well-being of 87; *see also* longtermism

Gates Foundation 14
Gibbard, Allan 202
give by earning 71, 93
GiveDirectly 93
GiveWell 9, 92, 93
Giving What We Can 96
global justice 12
global well-being *see* well-being
good: impartial good 131, 146–8; *see also* greater good; most good (doing)
good lives (value of) 124–6
government aid 32
governments: failure of 31–2; injustices of 28, 37–8, 45–7; moral responsibility of 46; political power of 52
government spending 31
gratitude (moral view of) 164–5, 188–9, 197, 209–10; benevolent motives 198; drowning child example 188–9, 199; versus elevation 210; supererogatory actions 139, 182, 189, 209
greater good 94
Greene, Joshua: "*The Psychology of (In)Effective Altruism*" 95

Hampton, Jean 194
Hanson, Robin: *The Elephant in the Brain* 77–8

help (moral requirement to) *see* moral requirement (to help)
Henderson, Rebecca 20, 56
hermit example (duty of beneficence) 109–10
Hidalgo, Javier 12
historial injustice 11, 44
Huemer, Michael 12
humanity (existential risk to) 125–6, 133–5, 157–8
"hyperagency" (of elite philanthropy) 15–16

immigration 12
impartial benevolence 156
impartial good 131, 146–8, 150, 152
impartial value 149–50
impermissibility 90
income (unequal distribution of) 11–12
influence: of corporate activism 56; within corporate environment 34; cultural 31, 39; of elite philanthropists 21–2, 38–9, 54; *see also* veto (power of)
injustice: distributive 11–12; everyday complicity 10, 28; fighting (as ethical action) 116–17; historical 11, 44; and immigration 12; inaction 28; of state (complicity in) 28, 37–8, 45–7, 51–2; wealth through 43; *see also* justice
intentions: judgment of 73; malicious 155; morality of 120
"interpersonal warmth" 147–8
intuition: human 168–9; moral 208

justice: domestic (in United States) 11–12; as donor constraint (bank robbery example) 9–10, 36–7, 44; global 12; reparative 36; *see also* injustice
"Just War Theory" 12

Kantian perspectives (of EA) 98

landmine example (future generations) 141
Lechterman, Theodore 17
Le Guin, Ursula: *The Ones Who Walk Away from Omelas* 195
liberals 11
libertarians 11, 36
Liberto, Hallie 173
liberty (moral) *see* moral liberty
lifeguard example 11, 37, 44–5
lives: meaningful 193; moral *see* moral lives
longtermism 126–7, 132, 157; consequentialist view of 132–5; and self-sacrifice 157–8; tiny probabilities 157–8, 159n7
looking good versus doing good 76–9, 101, 121, 129, 145

MacAskill, William 126; *Doing Good Better* 27, 64
McFarquhar, Larissa: *Drowning Strangers* 148
market failure 74, 106
meaningful lives *see* lives
meditation 185
Mill, J. S. 202
Minimally Effective Altruism *see* Effective Altruism (EA)
moral actions 120, 130, 139, 185
moral agents: assessment of 117, 130–1, 146, 155–6; as curse 194; *see also* virtue signals
moral assessment (of beneficence) 118–19
moral complacency 200–1
moral fanaticism 202
moral freedom (and donor discretion) 8–9, 163
moral intentions 120
moral intuition 208
morality: commonsense morality 66–7; demanding versus minimalist view 208–9; impartial 207–8; moral utility curve 137, 139; rules of 167, 209

moral liberty 173, 184, 200–1
moral lives 184–6; and Effective Altruism (EA) 192, 203, 213–14; good and bad 191–2
moral motivation 95, 101–2, 120–2, 140, 153
moral opportunity cost 8
moral prioritization (of Effective Altruism) 115
moral reasons 119–20; for beneficence 98–9, 101, 115, 153–4
moral requirement (to help) 166–9, 183–4, 199
moral risks: cause prioritization 201–2; of Effective Altruism (EA) 169–73, 183–4, 190, 200, 211–13; financial advisor example 212–13
moral theories 66, 89, 90–1, 130
moral utility curve 137, 139
most good (doing) 63, 68, 93, 99, 118–19, 121
motivations 76
Musk, Elon 31

Nagel, Thomas 185, 193
neutrality 41–2
noninterference (argument from) 166–7, 198, 210–11

objective concerns 203–5
obligation (moral) 164–8, 190; and gratitude 164–5, 188–9, 197; to help 66–7, 71, 82, 100, 109–11, 163–9
Oliver, Mary 100
"open-borders" 12
opportunity costs 8, 122, 131
oppression 42
ordinary philanthropy 13–14
overpoliticization (corporate activism) 23–4, 41, 47

Parfit, Derek: *Future-Tuesday Indifferent* agent 203–4; "Mere Addition Paradox" 125
pathological altruism 149, 194
Paul, Darel 20

"permanent stagnation" 135
permissibility 90
person-affecting views 123–4
personal consumption (frivolous) 8
philanthropic vouchers 39–40
philanthropy: Default View of 7–8, 13, 43; and democracy 39–41, 54, 170; and Effective Altruism (EA) 27; elite 14–16; inefficient 8; moral 13; ordinary 13–14; publicizing 120–1, 129, 140; Reich's definition of 7
philosophical theories (of beneficence) 187–8
physical body (rights to) 163, 173
Plummer, Theron 67–8
plutocracy: versus democracy 16, 54–5, 57; and elite philanthropy 15–16, 18
Pogge, Thomas 12
political equality: and corporate activism 21–3, 25; and elite philanthropy 17–18, 53, 55; veto (power of) 22, 55–7
political neutrality 41–2
political power 52
political saturation 24
political signalling 24
political values (versus beneficence) 49–51
politics: and corporate activism 33–4, 41; of universities 33
Pollock, John 142
pool club (lifeguard) example *see* lifeguard example
population ethics 125–6
poverty (global) 74–5, 92–3; aid illusion 108; eliminating 107–8; importance of development 108–9; injustice of 12; role of charities 109; *see also* prosperity (global)
"poverty porn" 80
prioritarians 87

priorities (objective versus subjective) 203–5
procedural neutrality 41–2
procedural unfairness (of elite philanthropy) 15, 38, 54
profitability 85–6
prosperity (global) 84; *see also* poverty (global)
psychological egoism 78
public opinion 22–3
Pummer, Theron 27

radicals 11
Reich, Robert 7
religion 191–2
reparative justice 36
rights: and Effective Altruism (EA) 171; of others, respect for 180–1, 202; to physical body 163, 166, 173; violations of 170, 180

sacrifice: and beneficence 116, 140; commonsense morality 67, 73; and longtermism 157–8; utilitarianism 64; valorization of 120
salience 90, 94
Samuel, Sigal 192
sandwich example (commonsense morality) 67
Saunders-Hasting 15, 17
saving lives (importance of) 8, 116–17
Schervish, Paul 15
Schmidtz, David 73, 138, 140
Schubert, Stefan: *"The Psychology of (In)Effective Altruism"* 95
"scope insensitivity" 77
self-care 194–5
selfishness 78; political 199
selflessness (as virtue signal) 123
self-sacrifice *see* sacrifice
Setiya, Kieren 167
signaling 75–6, 97; virtue signals 119–22
silent giving 121

Simler, Kevin: *The Elephant in the Brain* 77–8
Singer, Peter 8, 13, 75, 82, 92, 121, 139
skills (best use of) 122–3
Smart, J. C. C. 147
social businesses 69–70, 83, 93
social desirability bias 101, 208
social good 85–6
social norms, shifting 96
South-East Asia (economies) 74–5, 91–2
stakeholder capitalism 19
states *see* governments
subjective concerns 203–5
substantive neutrality 41–2
supererogatory actions *see* actions

Talisse, Robert 24
tax-payers (complicity of) 29
Temkin, Larry 9, 13, 51
Torres, Émile P. 125–6, 134–5
Tosi, Julian 190
total utilitarianism *see* utilitarianism
Tyson Foods 86

undemocratic: versus democratic 30–1; Effective Altruism (EA) as 170
Unger, Peter 139
unintentional beneficence 94
United States: charity donations 79; domestic justice 11–12; global injustice 12; immigration 12; "Just War Theory" 12; Religious Freedom Restoration Act (2015, Indiana) 22; US government 32; "war on drugs" 12
universities (political stance of) 33
utilitarianism 82; and beneficence 91, 115, 138; concern for future generations 123–4; defining 63, 66; and Effective Altruism (EA) 63–4, 66, 89, 98, 170–1; features of 89; and moral fanaticism 202; moral perspective of 89; partial commitments 80n3; on pathological altruism 194; total 123–4

value creation 86
values (moral): beneficence 116, 137–9; impartial 149–50
veto (power of) 22, 55–7
virtue: acting virtuously 121–2, 140, 156; benevolence as master virtue 131–2, 155; evidence of 119
virtue signals 119–22, 129, 140; selflessness 123; volunteering 122
virtuous agents 121–2, 140, 149, 156
volunteering: motives for 76–9; opportunity costs of 122; school example 182
"voluntourism" 76
voters: complicity of 29; democratic voting 32

war 12
Warmke, Brandon 190
wealth: redistribution of 39–40, 93; through injustice 43; unequal distribution of 11–12
well-being: and altruism 149, 163; of animals 86; and capitalism 84, 88n4; of future generations 87, 123–7, 132–5, 141–3; global/overall 90, 98, 115–17
"woke capitalism" 20
work (beneficence of) 72–4, 84, 100
workers (versus corporate interests) 34

Zuckerberg, Mark 182–3; donation to Newark schools 14–17, 31, 54, 211–12

Printed in the United States
by Baker & Taylor Publisher Services